10th International Conference on Environment and Renewable Energy (ICERE 2024)

E3S Web of Conferences Volume 514 (2024)

Nha Trang, Vietnam
25 – 27 February 2024

Editor:

P.-C. Chiang

ISBN: 978-1-7138-9549-7

Printed from e-media with permission by:

Curran Associates, Inc.
57 Morehouse Lane
Red Hook, NY 12571

Some format issues inherent in the e-media version may also appear in this print version.

For additional information, please contact EDP Sciences – Web of Conferences
at the address below.

EDP Sciences – Web of Conferences
17, Avenue du Hoggar
Parc d'Activité de Courtabœuf
BP 112
F-91944 Les Ulis Cedex A
France

Phone: +33 (0) 1 69 18 75 75
Fax: +33 (0) 1 69 28 84 91

contact-edps@webofconferences.org

Additional copies of this publication are available from:

Curran Associates, Inc.
57 Morehouse Lane
Red Hook, NY 12571 USA
Phone: 845-758-0400
Fax: 845-758-2633
Email: curran@proceedings.com
Web: www.proceedings.com

TABLE OF CONTENTS

ECOLOGICAL ENVIRONMENT MONITORING AND POLLUTION CONTROL

SOLID WASTE MANAGEMENT AND WASTE UTILIZATION

RENEWABLE ENERGY TECHNOLOGY AND ENERGY MANAGEMENT

GREEN BUILDING MATERIALS AND INDOOR AIR QUALITY

2024 10th International Conference on Environment and Renewable Energy

(ICERE 2024)

February 25-27, 2024, Nha Trang, Vietnam

COMMITTEE

Statement of Peer review

In submitting conference proceedings to *Web of Conferences*, the editors of the proceedings certify to the Publisher that

1. They adhere to its **Policy on Publishing Integrity** in order to safeguard good scientific practice in publishing.
2. All articles have been subjected to peer review administered by the proceedings editors.
3. Reviews have been conducted by expert referees, who have been requested to provide unbiased and constructive comments aimed, whenever possible, at improving the work.
4. Proceedings editors have taken all reasonable steps to ensure the quality of the materials they publish and their decision to accept or reject a paper for publication has been based only on the merits of the work and the relevance to the journal.

Title, date and place of the conference 2024 10th International Conference on Environment and Renewable Energy

February 25-27, 2024, Nha Trang, Vietnam

Proceedings editor(s): Prof. Pen-Chi Chiang

Date and editor's signature

2024 10th International Conference on Environment and Renewable Energy (ICERE 2024)

February 25-27, 2024, Nha Trang, Vietnam

PREFACE

This proceeding includes original and peer-reviewed research papers from 2024 10th International Conference on Environment and Renewable Energy (ICERE 2024) that was held during February 25-27, 2024 in Nha Trang, Vietnam.

We were honored to invite academic elites from all over the world to discuss the latest advances and cutting-edge research in Environment and Renewable Energy. Dr. Phan Phien and Dr. Phan Duc Ngai from University of Khanh Hoa, Vietnam were invited to do opening remarks for the conference. Prof. Marco Abbiati from University of Bologna, Italy; Prof. Pen-Chi Chiang from National Taiwan University, Taiwan; Prof. Dimitrios Karamanis from University of Patras, Greece have been invited to given keynote speeches. Prof. Daniel Ambach from Digital Business University of Applied Sciences, Germany; Prof. Tseung-Yuen Tseng from National Yang Ming Chiao Tung University, Taiwan delivered invited speeches at the conference. We believe that these reports have provided participants with insight and inspiration, and promoted research and development in related fields. It was an ideal platform for participants to exchange and interact with each other, helping to share experiences, promote cooperation and build new partnerships.

All accepted papers presented at the ICERE 2024 were included in this volume, that contained four chapters with topics: (1) Ecological Environment Monitoring and Pollution Control. (2) Solid Waste Management and Waste Utilization. (3) Green Building Materials and Thermal Comfort. (4) Renewable Energy Technology and Energy Management. All papers were subjected to peer-review by conference committee members and international reviewers.

At last, we strongly believe that the Proceedings of ICERE 2024 Conference will be of great significance to the academic community. We would like to present our heartfelt thanks and deeply appreciated to all of technical and conference committee members including colleagues who have organized the conference successfully. We are looking forward to having your active participation in future series conference of ICERE.

Prof. Pen-Chi Chiang

National Taiwan University, Taiwan
March 13, 2024

Assessment of Arsenic (As) and Lead (Pb) Concentration in the Water and Sediments of the Angat River Network in Banga II, Plaridel, Bulacan

Nanette Santos[1,2*] *Lhian Maxene* Venturina[1,2] and *Mary Abelene* Violago[1,2]

[1]School of Chemical, Biological, and Materials Engineering and Sciences
[2]Mapúa University, Intramuros, Manila, Philippines

Abstract. The Angat River Basin serves as a power and water source for Manila. However, the urbanization, industrialization and development in the surrounding areas could affect the water quality of the river. This study assessed the concentrations of arsenic and lead in the Angat River Network in Banga II, Plaridel, Bulacan, using Atomic Absorption Spectroscopy (AAS) and colorimetry from January to March 2023. All samples were examined at F.A.S.T. Laboratories in Cubao, Quezon City. Throughout the three months, the results of the water samples remained constant and were within permissible limits set by WHO, EPA, and DENR. On the other hand, the sediment samples show monthly variations in their results, but the concentrations remained within the permissible limits set by WHO and WSRA. Based on the findings, it can be concluded that the Angat River Network in Banga II, Plaridel, Bulacan, as classified by the DENR, still meets the safety standards of Class B/C, which is suitable for its intended usage after appropriate treatment.

1 Introduction

Water is essential for all life on Earth. Clean, fresh water is necessary for drinking and sanitation, providing for our crops, livestock, and industry, and creating and sustaining the ecosystems on which all life depends [1]. However, in today's world, water has become a significant threat. Water contamination by heavy metals is a serious environmental hazard that adversely affects all creatures exposed to it, including humans, animals, and plants. As

*Corresponding author: ndsantos@mapua.edu.ph

the world's population rises, development and industrialization rapidly become the primary cause of water contamination [2].

The heavy metal contamination of river waters in the Philippines directly results from industrialization and development [3]. Heavy metals are metals that are harmful at low concentrations and can have an impact on and threaten human health. Arsenic, chromium, copper, cadmium, mercury, lead, and nickel are some of the rivers' most prevalent heavy metals. [4]. The Angat River in the province of Bulacan comprises three dams: the Angat Dam, the Ipo Dam, which provides drinking water to Metro Manila residents, and the Bustos Dam, which feeds water to crop areas. The river has a basin area of 1,085 square kilometers and flows from the Sierra Madre highlands to Manila Bay. Angat River surrounds eleven municipalities in the province of Bulacan, including Angat, Baliwag, Bustos, Calumpit, Doña Remedios Trinidad, Hagonoy, Norzagaray, Paombong, Plaridel, Pulilan, and San Rafael. The Angat River Network in Plaridel, Bulacan, is part of the Angat River's midstream [3].

Heavy metal pollution in the river affects aquatic organisms, including the benthic and pelagic aquatic creatures that may suffer from sediment pollution or the aquatic organisms by causing severe damage that endangers their health and would severely impact aquatic ecosystems. Most heavy metals can also harm humans and pollute the ecosystem and atmosphere. Based on the water quality Status Report for Region 3 from 2001 to 2005, heavy metals such as arsenic, cadmium, chromium, copper, iron, lead, manganese, mercury, and zinc are present in the Angat River and are within DENR limits [5]. Additionally, according to the Environmental Monitoring Report of Angat River from July to October 2020, the presence of lead and arsenic were detected in the river but were within permissible limits [6].

Despite the presence of the said heavy metals being within permissible limits in previous studies, this could have changed throughout the years due to the rapid urbanization and other factors affecting the river such as weathering of soils and rocks from volcanic eruptions, from varieties of anthropogenic activities that involve mining, agricultural and industrial activities, combustion of fossil fuel and gasoline, waste incinerators, processing and use of metals and substances that contain metal pollutants, etc. There are multiple studies about the concentration of heavy metals in the Angat River. However, this study will focus on assessing the arsenic and lead concentrations in the water and sediments from the Angat River, specifically, Banga II, Plaridel, making it a novel contribution to existing research studies.

The river in the municipality of Plaridel is bordered by agricultural, residential, transportation, and commercial establishments. There are also industrial establishments within Banga II and the neighbouring communities. The river also provides food and livelihood to the residents who live nearby. The waste emissions from these operations may harm the river, aquatic species, and residents who rely on the river for a living. Upon ocular inspection of the area, the grey colour of the water can be vividly seen, and the riverbanks are full of solid wastes, especially in the residential sector. Therefore, the Angat River Network's quality must be assessed considering recent activities in Plaridel, Bulacan. The potential for the river to be contaminated with lead, arsenic, or any other heavy metal is a threat to the residents and to the aquatic species living in the river.

The objective of this study was to determine whether the concentrations of lead and arsenic in water and sediment samples collected from the Angat River in Plaridel, Bulacan, meet the acceptable limits established by various organizations such as the World Health Organization (WHO), Environmental Protection Agency (EPA), Department of Environment and Natural Resources (DENR), and World Surface Rock Average (WSRA). The results of water samples were compared to the guidelines set by WHO, EPA, and DENR. The provisional guideline value provided by WHO for both arsenic and lead is 0.01 mg/L [7]. The EPA recommends a value of 0.01 mg/L for arsenic and 0.015 mg/L for lead [8]. The DENR's recommended value for both arsenic and lead, based on Class B/C classification, is

0.05 mg/L [9]. For sediment samples, the results were compared to the guidelines provided by WHO and WSRA. WHO's provisional guideline specifies a limit of 20 mg/kg for arsenic and 100 mg/kg for lead in sediment. WSRA recommends a 10 mg/kg limit for arsenic and 16 mg/kg for lead [10].

The collected samples undergo testing at First Analytical Services and Technical Cooperative (F.A.S.T.) Laboratories to determine the levels of heavy metals, specifically lead and arsenic, present in the Angat River. Water and sediment samples were tested using Flame Atomic Absorption Spectrometry (AAS) for lead analysis. Arsenic analysis, on the other hand, was performed using colorimetry for both water and sediment samples. Flame Atomic Absorption Spectrometry (AAS) is a technique used to measure the quantity of chemical elements present in environmental samples by analysing the absorbed radiation emitted by the chemical component. It involves examining the spectra generated when the sample is stimulated by radiation. This method is precise, has low detection limits, requires simple sample preparation, and is cost-effective. It has proven an effective technique for analysing trace elements in various materials [11]. Colorimetry, on the other hand, is a scientific method employed to determine the concentration of coloured compounds in solutions. It utilizes the Beer-Lambert equation, which establishes a proportional relationship between the concentration of a solute and its absorbance. By applying this equation, colorimetry allows for the quantification of the concentration of coloured compounds in a sample [12]. Due to its simplicity, ease of use, and speedy on-site detection, colorimetry is one of the most popular techniques for detecting heavy metal ions [13].

Monitoring of potentially toxic metals in water and sediments is essential as it provides valuable information about the water's overall quality and helps in assessing the environmental risk posed by these metals [14]. This study presents information about the heavy metal content in the Angat River. This river serves as a habitat for aquatic organisms and a vital water source for residential and agricultural sectors, aligning with the sustainable development goals 14 and 15, which focus on life below water and life on land, respectively. According to the sustainable development goal 6 which emphasizes clean water and sanitation, the water quality remains unknown for about 3 billion individuals due to insufficient monitoring. By 2030, the goal is to enhance water management by encouraging local communities to play a significant role in monitoring and improving water management practices [15]. The results of this study serve as an initial evaluation of the river's toxicity level due to arsenic and lead concentrations, providing the municipality with valuable information for planning actions to enhance the river's water quality. The municipality was informed about the study and agreed that it would be shown to them upon getting the results, if the concentrations exceed the established limits, the municipality will be responsible for formulating and implementing the necessary actions to address the situation.

The study's scope and delimitations are outlined as follows: (1) The collection and testing of water and sediment samples is limited to a single sampling point. (2) Samples were gathered and tested once every month. (3) The analysis only focused on lead and arsenic as heavy metals of interest. (4) The upper and lower streams of the river were not considered, as the study pertains only to the Plaridel Network of the Angat River, which represents the middle stream section. Due to these scopes and limitations, the study's findings primarily apply to the area of the river where the samples were collected. The concentration of heavy metals can vary across different regions of the river, influenced by various factors and surroundings. Moreover, it is essential to note that this study will not encompass water quality monitoring parameters such as faecal coliform count, total coliform count, pH level, turbidity, temperature, electrical conductivity, dissolved oxygen (DO), biological oxygen demand (BOD), chemical oxygen demand (COD), total organic carbon (TOC), total dissolved solids

(TDS), and total suspended solids (TSS). These aspects are beyond the scope of the current study.

2 Methodology

2.1 Research locale

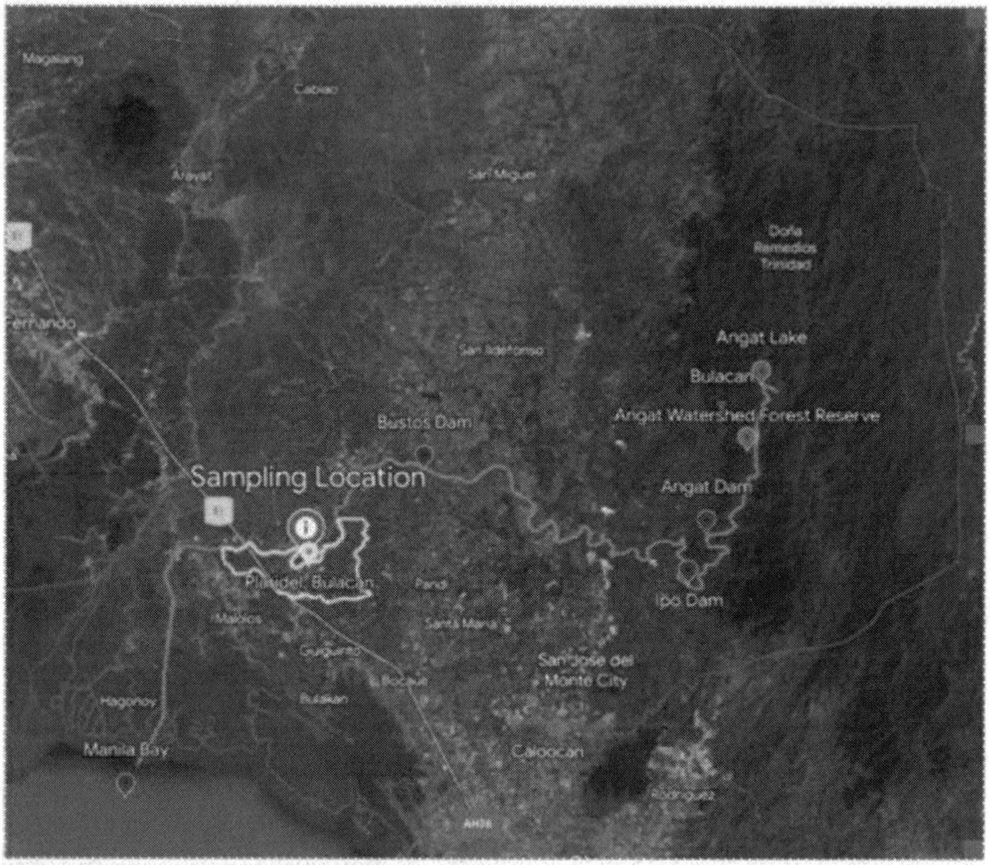

Fig. 1. Angat River Stretch in Bulacan, Province

The Angat River, the largest river in Bulacan, flows from the Sierra Madre Mountain range to Manila Bay. Figure 1 illustrates the Angat River stretch in the province of Bulacan to Manila Bay, represented by the blue line, while the yellow border line represents the municipality of Plaridel, Bulacan. The river meanders through numerous municipalities in Bulacan, including Doña Remedios Trinidad, Norzagaray, Angat, Bustos, San Rafael, Baliwag, Plaridel, Pulilan, Calumpit, Paombong, and Hagonoy. The river joins the Pampanga River through the Bagbag River in Calumpit, Bulacan [16]. The river encompasses an area of 1,085 square kilometers within the Angat Watershed Forest Reserve. Several dams are situated along the Angat River, namely the Angat Dam, Ipo Dam, and Bustos Dam. The Angat Dam alone provides a daily supply of 500 million gallons of water, meeting 97% of the water needs of Metro Manila [17].

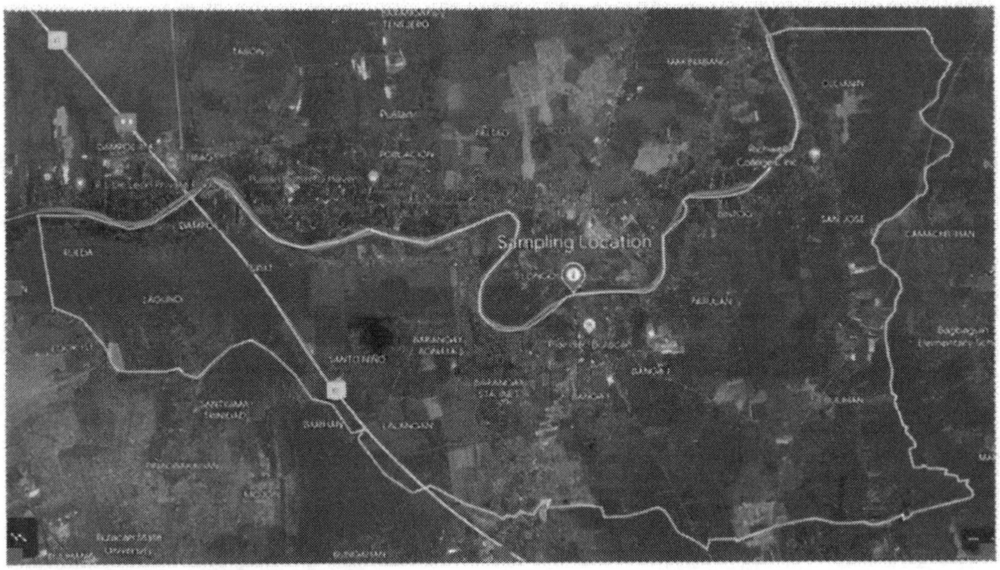

Fig. 2. Municipality of Plaridel, Bulacan

Figure 2 illustrates the municipality of Plaridel, Bulacan, which consists of 19 barangays, namely Agnaya, Bagong Silang, Banga I, Banga II, Bintog, Bulihan, Culianin, Dampol, Lagundi, Lalangan, Lumang Bayan, Parulan, Poblacion, Rueda, San Jose, Sta. Ines, Santo Nino, Sipat, and Tabang. Due to the vast expanse of the Angat River, this study specifically focuses on the section of the river located in Banga II, Plaridel, Bulacan. This section represents the middle stream and is chosen due to its convenient accessibility for sample collection. The researchers will focus on one sampling point in this section to gather water and sediment samples. The sampling point has a latitude of 14°53'29" N and a longitude of 120°52'0" E. According to Comprehensive Land Use Plans (CLUP), Banga II belongs to the identified central business district in Plaridel, which is why the area is surrounded by commercial and industrial establishments, agricultural, transportation, and residential areas. The river's surroundings were considered when selecting the sampling point because numerous factors can contribute to the heavy metals in the river.

All water and sediment samples collected from the river undergo testing at the F.AS.T. Laboratories. Skilled technical experts specializing in testing and operational areas performed the testing procedures. The laboratory strictly follows internationally recognized testing methodologies and employs approved instrumentation. Additionally, it has been accredited and recognized by multiple government agencies.

2.2 Research design

This research is an experimental study that involves quantitative data analysis. The study aims to assess the concentration of lead and arsenic present in both water and sediments within the Angat River Network, specifically in Banga II, Plaridel, Bulacan. The analysis will use atomic absorption spectrometry (AAS) and a colorimetric method. The study will span three months, from January 2023 to March 2023, considering factors such as time, weather conditions, and the surroundings of the sampling location.

2.3 Sample sourcing

2.3.1 Water sample sourcing

To collect the water sample, a polyethylene terephthalate bottle with a cap with a 1L volume was employed using the grab sampling method. A glove was worn before the collection of the water sample. Facing toward the direction of the water flow, the initial step involved rinsing the PET bottle. Removing the bottle cap, the container was inverted and submerged below the water surface, ensuring that at least one-third of the container volume was filled with river water. Gently shaking the container, the water was poured downstream from the collected sample. This rinsing process was repeated three times. To obtain the sample, the container was again submerged below the water surface until it was filled. The container was labelled with the date, time, and weather conditions.

2.3.2 Sediment sample sourcing

A 4L polypropylene bucket pail and a stainless-steel scoop were employed to collect sediment samples. While wearing a glove, grab samples of the sediment were scooped from the sampling point where the water sample was obtained. The container was labelled with the date, time, and weather conditions.

2.4 Sample preparation

To preserve the quality of the samples, they were transported to the F.A.S.T laboratory within six hours from the moment of collection under ambient temperature. This prompt transport was done to minimize the potential changes in the chemical parameters of the samples. The professionals conducted all the testing procedures in accordance with the standardized methods specified in the 23rd edition of APHA AWWA and WEF 2017 for examining water and wastewater, as well as the 3rd edition of USEPA's test method for evaluating solid wastes.

2.5 Analysis of heavy metal concentration in the samples

Lead concentrations in the water and sediment samples were determined using 3111 B Direct Air Acetylene Flame Atomic Absorption Spectroscopy (AAS). For arsenic analysis in the water sample, the approach used was the 3500-As B method, specifically the silver Diethyldithiocarbamate method. This method is a colorimetric technique designed to determine arsenic in aqueous samples. As for the sediment sample, a modified colorimetric method was employed to analyse arsenic concentrations.

3 Results and discussion

In assessing the concentrations of lead and arsenic in the water and sediment samples from the Angat River, various factors were considered, including time, wind, weather conditions, and the surroundings of the sampling point. To account for the time, the researchers opted to collect samples during mid-morning when businesses are operational, which increases the likelihood of waste movement that may contain metals from these establishments. The sampling point's surroundings include a slaughterhouse, a resort, a mall, residential areas, and transportation routes. Considering these parameters is crucial, as they can affect the river's water quality.

During January and March, the wind was characterized as a light breeze, and the weather was rainy two days before the sample gathering, followed by a sunny day during the sample collection. In February, the wind was described as a moderate breeze that may carry dust, leaves, and paper. The weather was rainy two days before the sample gathering, with a cloudy day during the sample collection. Accounting for wind is also essential, as it can introduce airborne pollutants into the water. Ahmed T. et al. state that storm events, snowmelt, drought, and elevated air temperatures significantly impact drinking water quality. For example, heavy rainfall can increase water turbidity, while temperature changes can affect the chemical quality of water [18].

Table 1. Heavy Metal Concentration in Water Sample

Pollutant	**January (mg/L)**	**February (mg/L)**	**March (mg/L)**	**WHO (mg/L)**	**EPA (mg/L)**	**DENR (mg/L)**
Arsenic (As)	< 0.005**	< 0.005**	< 0.005**	0.01	0.01	0.05
Lead (Pb)	<0.01*	<0.01**	<0.01**	0.01	0.015	0.05

Table 1 presents the test results concerning the water samples obtained from the Angat River in Banga II, Plaridel, Bulacan. The results indicated that over three months, the lead concentration was less than 0.01** mg/L, while the concentration of arsenic was less than 0.005** mg/L. Notably, both heavy metals' concentrations fall within the acceptable limits established by WHO and EPA for drinking water. Furthermore, these concentrations adhere to the allowable limits the DENR set for water quality criteria for Class B/C concerning Toxic and Other Deleterious Substances in Fresh Waters.

Table 2. Heavy Metal Concentration in Sediment Sample

Pollutant	**January (mg/kg)**	**February (mg/kg)**	**March (mg/kg)**	**WHO (mg/kg)**	**WSRA (mg/kg)**
Arsenic (As)	< 0.005**	0.02	< 0.01**	20.0	10.0
Lead (Pb)	<0.01**	0.04	<0.005**	100.0	16.0

The concentration of the heavy metals, arsenic, and lead, determined in the sediment samples collected from the Angat River, are presented in Table 2. The arsenic concentrations were < 0.005**, 0.02, <0.01** mg/kg for January, February, and March, respectively. Similarly, the concentrations of lead were <0.01**, 0.04, <0.005** mg/kg for the corresponding months. Based on the results, the detectable levels of both heavy metals emerged in February, possibly influenced by the weather conditions and the ongoing construction near the sampling site, which started in February and stopped in March. Construction materials such as welding or paints, have the potential to release heavy metals like arsenic, lead, and chromium [19]. Moreover, numerous studies have looked at the relationship between sediment grain size and heavy metal content. The major finding is that metal concentration rises as grain size decreases [20]. However, this is only applicable if it is naturally available and not artificially included. Nonetheless, the concentrations of both metals remain within the acceptable limits established by the WHO and WSRA for sediment samples.

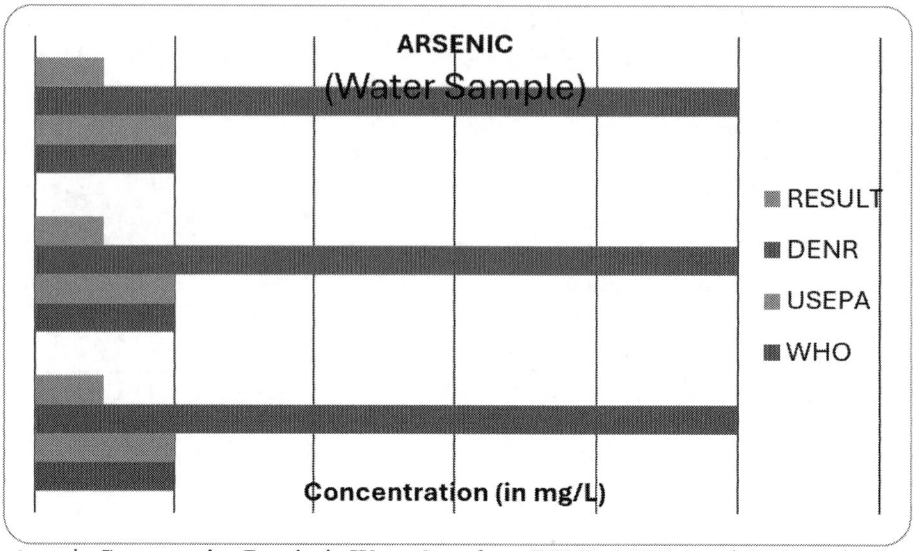

Fig. 3. Arsenic Concentration Results in Water Sample vs DENR, USEPA and WHO Guideline Values

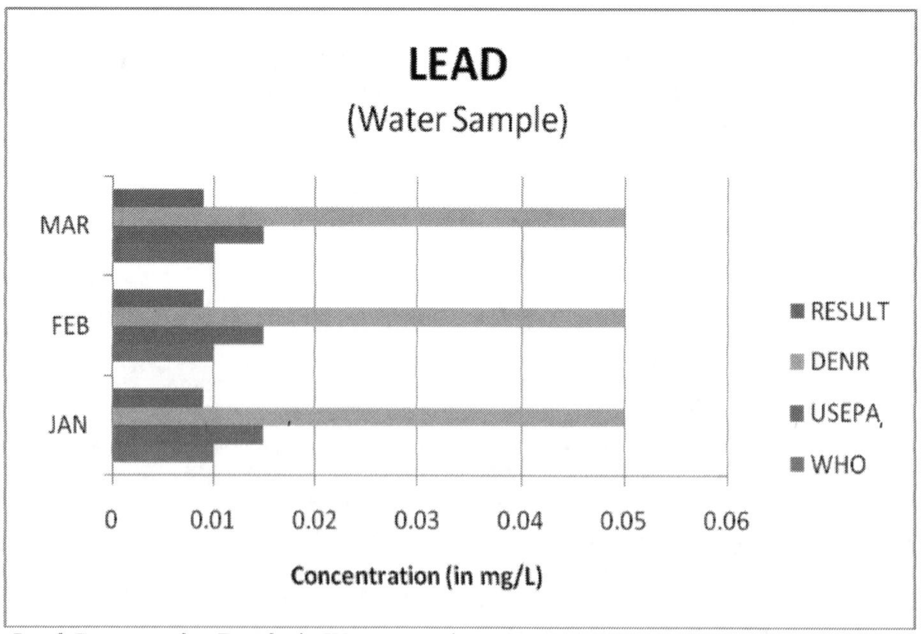

Fig. 4. Lead Concentration Results in Water Sample vs DENR, USEPA and WHO Guideline Values

 https://doi.org/10.1051/e3sconf/202451401001

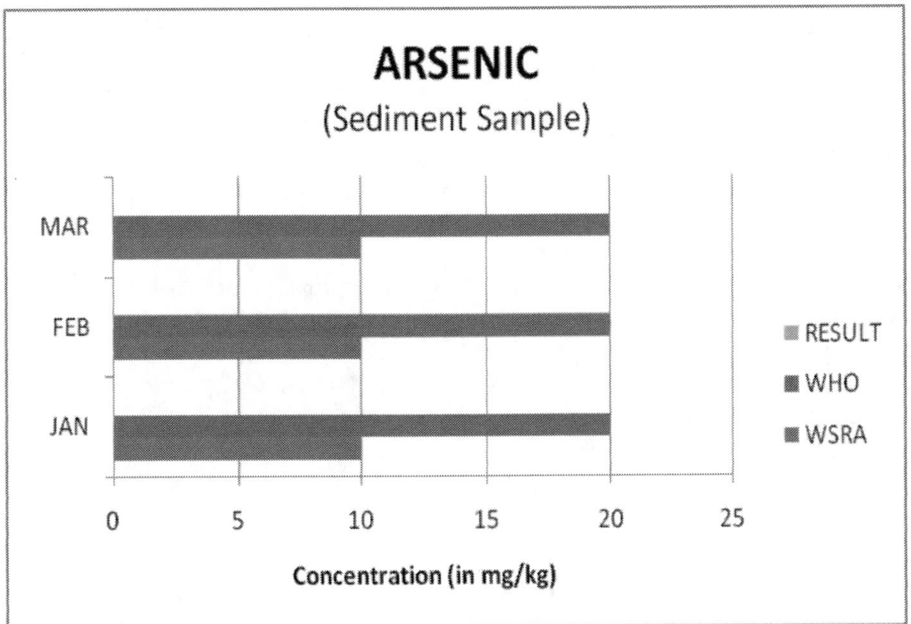

Fig. 5. Arsenic Concentration Results in Sediment Sample vs WHO and WSRA Guideline Values

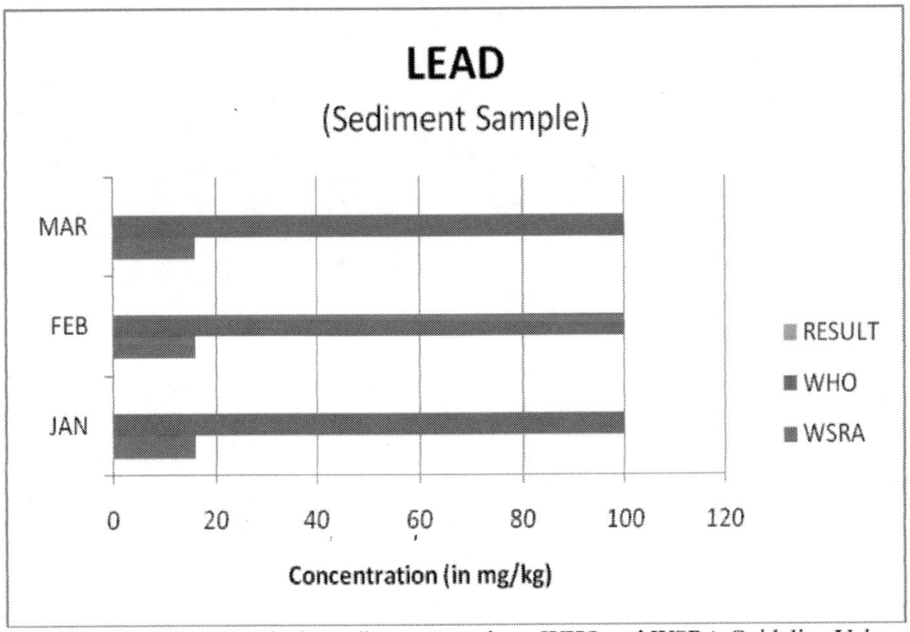

Fig. 6. Lead Concentration Results in Sediment Sample vs WHO and WSRA Guideline Values

Figures 3-6 represent the graphical presentation of the permissible limits and concentrations of lead and arsenic obtained in the water and sediment samples. According to F.A.S.T Laboratory, when the result falls below the minimum detection limit, it will not be assigned a specific value. This clarifies why no precise values are recorded in the results obtained from the water and sediment samples.

Given that the Angat River serves as a vital water source for both humans and aquatic organisms, this research study suggests conducting tests on other physiochemical and biological parameters which are key factors affecting water quality such as pH, turbidity, dissolved oxygen (DO), biochemical oxygen demand (BOD), total suspended solids (TSS), total dissolved solids (TDS), fecal coliform and total coliform for a more comprehensive understanding of its overall water quality [21]. To better understand the impact of weather variations and climatic zones, it is advisable to extend the study duration. This would allow for a more extensive examination of weather patterns in heavy metal concentrations [22]. Since this study only conducted one trial per month, it is recommended that future studies increase the number of sampling trials for more comparison of heavy metal concentrations in the Angat River.

Furthermore, this study focused on a single sampling point, specifically Banga II, Plaridel, Bulacan. Testing multiple sections of the Angat River is advisable to enhance understanding and obtain more accurate results. Surrounding factors such as the urban and residential areas have an influence in the river and may impact heavy metal concentrations [23]. Testing different areas would provide a more comprehensive picture of the overall water quality of the river.

4 Conclusion

The study utilized Atomic Absorption Spectroscopy (AAS) and Colorimetry to assess the levels of lead and arsenic in the Angat River Network located in Banga II, Plaridel Bulacan, over three months. The aim was to determine if the concentration of the heavy metals is within permissible limits over the years. The findings indicate that the concentrations of lead and arsenic in both water and sediment samples from the Angat River Network are within the acceptable limits defined by various organizations such as WHO, EPA, DENR, and WSRA. However, it is essential to note that the concentrations observed in this area of the Angat River might differ from other parts. Therefore, it is recommended to conduct further testing to determine the levels of lead and arsenic in different sections of the Angat River.

Furthermore, the results obtained from this study still fall within the classification established by the DENR for Class B and C freshwater and remain appropriate for activities involving direct contact with water, including swimming, and for the growth and propagation of fish and other aquatic resources. It can also be utilized for activities such as boating and specific manufacturing processes after applying appropriate treatment measures.

References

1. UNEP – UN Environmental Programme. Why does water matter? (n.d.)
2. A. Singh, A. Sharma, R. Verma, et al. Heavy metal contamination of water and their toxic effect on living organisms. IntechOpen eBooks (2022)
3. E.A. Estrella, M.S. De Leon, E.N. De Regla, B.S.C. Duay, E.D.A. Estrella, Determination of lead concentration of the waters in the Angat River Network in Bustos, Bulacan, Int. J. ChemTech Res. 7 1895–1901 (2015)
4. Metals occurring in aquatic freshwater. (n.d.)
5. DENR (Department of Environment and Natural Resources), Region 3 Water Quality Status Report (RWQSR) for Public Information and Advocacy, 45 (2006)
6. M. Waterworks, S. System, Philippines: Angat Water Transmission Improvement Project (2014)
7. W.H. Organization, Chemical fact sheets, Guidel. Drink. Water Qual. 296–461 (2010)

8. H.N.M.E. Mahmud, A.K.O. Huq, R.B. Yahya. The removal of heavy metal ions from wastewater/aqueous solution using polypyrrole-based adsorbents: A review, RSC Adv. 6, 14778–14791 (2016)
9. Department of Environment and Natural Resources, PRIME - M4 Page 1 of 11, 1–11 (1990)
10. S.A. Onjefu, K.N. Anna, S.T. Halahala. Heavy Metal Seasonal Distribution in Shore Sediment Samples along the Coastline of Erongo Region, Western Namibia. The Effects of Atmospheric Particulate Matter on Climate and Human Health View project Heavy Metal Seasonal Distribution in Shore Sediment Sa, Eur. J. Sci. Res. 139, 49–63 (2016)
11. A.K. Jaiswal, M. Mohrana, P.H.R. Krishna, D. V. Moon, T. Millo, O.P. Murty, Atomic absorption spectrometry - A review, J. Forensic Med. Toxicol. 27, 64–7 (2010)
12. J. Lucas, P. Lucas, T.L. Mercier, A. Rollat, WG. Davenport. Applications of rare Earth luminescent materials. Elsevier eBooks. 281-318 (2015)
13. Z. Chen, Z. Zhang, J. Qi, J. You, J. Ma, L. Chen. Colorimetric detection of heavy metal ions with various chromogenic materials: Strategies and applications. Journal of hazardous materials, 441, 129889, (2023)
14. H. Ali, E. Khan, I. Ilahi. Environmental Chemistry and Ecotoxicology of Hazardous Heavy Metals: Environmental Persistence, Toxicity, and Bioaccumulation, J. Chem. 1–14 (2019)
15. United Nations, Goals @ Sdgs.Un.Org, United Nations Sustain. Dev. Goals. (2020)
16. G. Water, A. Life, E.N. De Regla, M.S. De Leon, E.A. Estrella, B.S.C. Duay, Determination of Ph Levels and Cadmium Content of Surface, 8, 1836–1845. (2015)
17. Asian Development Bank. Philippines: Angat Water Transmission Improvement Project. (2016)
18. T. Ahmed, M. Zounemat-Kermani, M. Scholz, Climate change, water quality and water-related challenges: A review with focus on Pakistan, Int. J. Environ. Res. Public Health. 17, 1–22, (2020)
19. D.M. Ceballos, Z. Dong, J.L. Peters, R.F. Herrick, P. Gupta, J.D. Spengler, Metals dust in workers' homes and potential for take home in the Greater Boston area: Pilot study, Environ. Res. 209, 112893, (2022)
20. E. Singovszka, N. Junakova, M. Balintova, The effect of sediment grain size on heavy metal content in different depth in water reservoir Ruzin, Slovakia, Solid State Phenom. 244, 240–245, (2016)
21. Nguyen Thanh Giao, "Determining Key Factors Affecting Surface Water Quality in Soc Trang Province, Vietnam," International Journal of Environmental Science and Development vol. 13, no. 5, pp. 144-149, (2022)
22. D. T. Jayawardana, H. M. T. G. A. Pitawala and H. Ishiga, "Groundwater Quality in Different Climatic Zones of Sri Lanka: Focus on the Occurrence of Fluoride," International Journal of Environmental Science and Development vol. 1, no. 3, pp. 244-250, (2010)
23. Nguyen Thanh Giao* and Nguyen Hong Thao Ly, "Influence of Urban and Residential Areas on Surface Water Quality in the Vietnamese Mekong Delta Coastal Province," International Journal of Environmental Science and Development vol. 14, no. 4, pp. 234-244, (2023)

E3S Web of Conferences **514**, 01002 (2024) https://doi.org/10.1051/e3sconf/202451401002
ICERE 2024

Optimization of step-feed anoxic-oxic-membrane bioreactor (AO-MBR) with methanol addition for biological nitrogen removal in wastewater using BioWin and response surface methodology (RSM)

Aika Cecille Tolentino[1,2], *Hai-Hsuan* Cheng[3], *Chih-Chieh* Lin[3], *Bonifacio* Doma, Jr.[1,2], and *Liang-Ming* Whang[3,4,*]

[1] School of Graduate Studies, Mapúa University, Manila 1002, Philippines
[2] School of Chemical, Biological, and Materials Engineering and Sciences, Mapúa University, Manila 1002, Philippines
[3] Department of Environmental Engineering, National Cheng Kung University, Tainan 70101, Taiwan
[4] Sustainable Environment Research Laboratory (SERL), National Cheng Kung University, Tainan 70101, Taiwan

Abstract. The study optimized a step-feed anoxic-oxic membrane bioreactor (AO-MBR) with methanol addition, an existing biological nitrogen removal (BNR) process, to improve total nitrogen (TN) removal and to reduce operating costs. Methanol, utilized as an external carbon source, was costly, which led to examining ways to reduce its usage by adjusting the operational parameters and modifying the configuration. The BioWin's built-in activated sludge model underwent calibration through biokinetic batch experiments and numerical adjustments of stoichiometric and kinetic parameters to create the BNR process model. Dynamic simulations validated the calibrated model. Using the model, a modified configuration with a 20.8% increase in total anoxic volume was proposed. This increased the anoxic hydraulic retention time, improving the denitrification process. The response surface methodology (RSM) was then used to optimize the operating parameters of the modified configuration for maximum TN removal efficiency and minimum operating cost. The optimized parameters found were 0.5 mg/L dissolved oxygen, 45 days of solid retention time, 0:1 methanol distribution ratio, and 1,794.29 L/d methanol, which all enhanced the denitrification process. This resulted in a 16% increase in TN removal efficiency, a 4.12% decrease in operating costs, and an 8.96% reduction in methanol usage.

1 Introduction

Excessive levels of nitrogen compounds discharged into water bodies may result in eutrophication, leading to various aquatic concerns and health problems. Hence, it is

* Corresponding author: whang@mail.ncku.edu.tw

necessary to treat such pollutants. Among the treatment processes for nitrogen, the activated sludge process is often utilized. This process uses microorganisms to remove nutrients and organic compounds from wastewater [1]. It involves complex elements sensitive to various factors and complicated systems comprising multivariate operations [2,3]. This complexity makes the successful operation of the activated sludge process a challenging task. Fortunately, mathematical modeling and simulation have been used over the years, and it is an effective tool for evaluating and predicting the treatment performance of a complex treatment process. Activated sludge models (ASMs) describe the biological reactions occurring within an activated sludge system by a series of differential equations [4]. It can provide an understanding of biological nitrogen removal (BNR) for efficient management, redesign, and optimization to attain desired goals (e.g., high treatment efficiency and cost reduction) [2,5,6].

In recent years, commercially available software incorporating ASMs, such as STOAT, WEST, and GPS-X, has been developed and used in municipal and industrial wastewater treatment plants (WWTPs) [4]. Among the simulation software, BioWin is widely utilized in research due to its high flexibility and ease of use [7]. BioWin is a wastewater treatment simulation tool that combines different model processes - physical, chemical, and biological [8]. The biological process model is a vital section of BioWin, where the built-in general model combines international models of activated sludge and anaerobic digestion [9]. It also integrates several models, such as sulfur, chemical precipitation, pH, alkalinity, and industrial component processes. It has more than eighty process expressions, including over fifty state variables. In the activated sludge alone, the BioWin built-in general model comprises the growth and decay of the following bacteria: ammonia-oxidizing bacteria, nitrite-oxidizing bacteria, heterotrophic bacteria, methylotrophic bacteria, anaerobic ammonia-oxidizing bacteria, and phosphorus-accumulating bacteria. It also contains hydrolysis, assimilative denitrification, ammonification, and the adsorption process of the microorganisms [8]. BioWin software has been utilized for optimization studies. An example is Elawwad et al. [10], who used BioWin's built-in ASM to calibrate and simulate the largest WWTP in Egypt. After they validated the model, a step-wise optimization of operating parameters was done to maximize the nitrogen removal.

Aside from step-wise optimization, response surface methodology (RSM) can be a tool in effective optimization studies for BNR simulations. RSM is an experimental design method that can evaluate the relationship between multivariate factors and responses [5]. With fewer experiments, empirical models can be formed, which can be utilized to analyze the effects of several factors and obtain the optimum conditions for responses. To our knowledge, only two studies so far have combined the ASM simulation with RSM for optimization [2,5].

In this study, a step-feed anoxic-oxic-membrane bioreactor (AO-MBR) with methanol addition, an existing BNR process, was simulated using BioWin's built-in ASM and optimized using RSM. The high cost of utilizing methanol as an external carbon source led to a concern in the current BNR process. Hence, this study explored ways to reduce methanol usage by adjusting operating parameters and slightly modifying the configuration. While the methanol cost was aimed to lower, the energy cost was also considered. The integration of RSM helped minimize the number of simulations for optimization. The operating parameters of the modified configuration were optimized for maximum TN removal efficiency and minimum operating cost. No study so far has explored the use of combined software (BioWin and RSM) for problem-solving cases on a step-feed AO-MBR with methanol addition.

2 Methodology

The step-feed anoxic-oxic-membrane bioreactor (AO-MBR) with methanol addition focused on this study belongs to the YK municipal wastewater treatment plant (WWTP) in Tainan, Taiwan. To simulate the said process effectively, the Good Modeling Practice (GMP) protocol by the GMP Task Group [11] was used as a guide. The protocol comprises five main steps: defining the project objective, collecting data and information, setting up the plant model, undertaking calibration and validation, and, lastly, simulating the model to achieve the goals.

2.1. YK municipal wastewater treatment plant

Information regarding the biological nitrogen removal (BNR) process of YK WWTP was provided by the engineer who works in that plant. Information such as unit configuration, design, and historical records on operations and performance were given.

The BNR process comprises seven tanks and a membrane bioreactor (MBR) with a total effective volume of 6,966.8 m^3. It is designed to have three stages of the Anoxic-Oxic (AO) process, where the influent splits and flows on each stage. The influent flow ratio is 2:3:1. Only the anoxic tanks are equipped with a mixer. The MBR replaced the settling tank for solid-liquid separation. It is fitted with submerged 27 hollow fiber PVDF membrane cassettes with a 0.04 μm nominal pore size –a total membrane surface area of 1,651.2 m^2/cassette. The average return-activated sludge concentration was 8,992.86 mg-SS/L. At the present configuration, the second tank was replaced from an oxic to an anoxic tank due to the high dissolved oxygen reflux from MBR. Also, to enhance denitrification, methanol is added and flows in equal amounts through the 2nd and 6th biological tanks. The operational parameters and volume of each tank are summarized in Table 1, and the flow chart of the BNR is presented in Figure 1.

Table 1. Operational parameters and volume of tanks.

Operational parameter	Values	Tanks	Volume [m^3]
Influent flowrate [m^3/d]	18,622.85 [a]	Anoxic Tank 1 (1st tank)	499.0
DO in Oxic tank 1 [mg/L]	0.5	Anoxic Tank 2 (2nd tank)	997.9
DO in Oxic tank 2 [mg/L]	1.0-1.5	Anoxic Tank 3 (3rd tank)	695.0
DO in MBR [mg/L]	3.5-5.0	Oxic Tank 1 (4th tank)	1,336.5
Solid retention time [day]	25-30	Anoxic Tank 4-1 (5th tank)	445.5
Total methanol (45%) flow [m^3/d]	1.97 [a]	Anoxic Tank 4-2 (6th tank)	445.5
Return-activated sludge flow [m^3/d]	36,149.23 [a]	Oxic Tank 2 (7th tank)	1,871.1
Waste sludge flow [m^3/d]	176.92 [a]	MBR	676.3

[a] Average values.

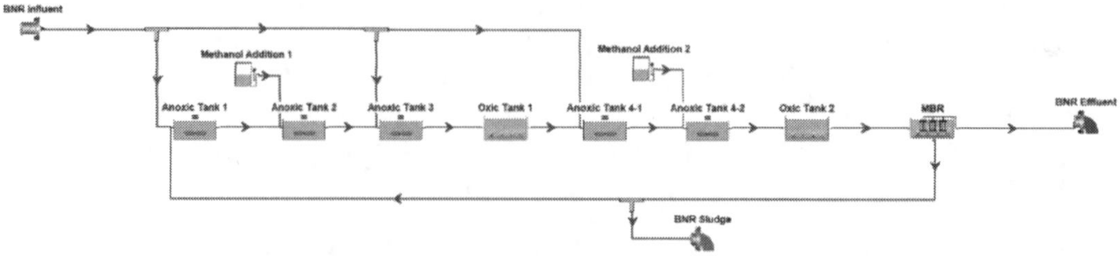

Fig. 1. BNR layout of the YK municipal WWTP (exported from BioWin software).

2.2. Wastewater characterization

For wastewater characterization, the wastewater quality of the influent BNR (wastewater after primary clarifier) and the effluent BNR (permeate flow from MBR) were analyzed for pH, total chemical oxygen demand (TCOD), suspended solids (SS), volatile suspended solids (VSS), ammonia-nitrogen (NH_4^+-N), nitrite-nitrogen (NO_2^--N), nitrate-nitrogen (NO_3^--N), total Kjeldahl nitrogen (TKN), total nitrogen (TN), and total phosphorus (TP). All analyses referred to the standard methods [12].

Part of wastewater characterization is identifying the influent COD fractions. BioWin's built-in activated sludge model (ASM) accounts for the fraction of influent COD. The TCOD was divided into four fractions: readily biodegradable COD (S_S), soluble unbiodegradable COD (S_I), slowly biodegradable COD (X_S), and particulate non-biodegradable COD (X_I), as shown in Equation (1).

$$TCOD = S_S + S_I + X_S + X_I \tag{1}$$

In this study, the S_I was considered equal to the soluble COD in the effluent [13]. S_S was obtained by subtracting the influent soluble COD from S_I. The influent soluble COD was determined using a physical-chemical method described by Mamais et al. [14]. X_I was computed by subtracting the TCOD of the 7th tank's supernatant by S_I. X_S was obtained as the remaining balance in Equation (1).

2.3. Biokinetic batch experiment

The built-in ASM model has more than eighty process expressions describing the biological process, including over fifty state variables [8]. Since many parameters are to be considered in modeling, this study referred to the sensitivity analysis conducted by Liwarska-Bizukojc and Biernacki [9] to identify the most influential parameter. The maximum specific growth rate of ammonia-oxidizing bacteria ($\mu max_{(AOB)}$), the maximum specific growth rate of heterotrophic bacteria ($\mu max_{(HET)}$), substrate (COD) half-saturation ($Ks_{(COD)}$), growth yield of heterotrophic under aerobic condition (Y_H), and aerobic decay rate (b_H) are some of the critical parameters that could influence the simulation and hence were selected to determine the actual values for YK BNR process.

Batch experiments were divided into three: nitrification experiment for $\mu max_{(AOB)}$ value; carbon addition experiment for $\mu max_{(HET)}$, $Ks_{(COD)}$, and Y_H values; and decay experiment for b_H value. All batch experiments were performed in a one-liter serum bottle(s) with aeration from air bubble stone. Influent wastewater was used as a substrate, and the sludge from the aerobic tank was used as inoculum. A magnetic stirrer continuously stirred the mixed liquor, and the pH was regulated between 7-8 by NaOH and H_2SO_4. Specifically, for the nitrification experiment, the MLVSS was diluted to 150 – 300 mg/L [15] For the carbon addition experiment, sodium acetate trihydrate was added as an external substrate to create varied initial concentrations of COD (200- 2,000 mg/L), thereby producing a different food-microorganism ratio (S_o/X_o) [13].

All batch experiment samples were taken over time. For the decay experiment, the sample was taken over days (lasting 15 – 20 days) [16]. The samples for nitrification and carbon addition experiments were filtered through a 0.45 μm membrane filter and analyzed for NO_2^--N and NO_3^--N, and COD, respectively. For the carbon addition and decay experiment, the VSS was monitored.

Following the research study by Antoniou et al. [15], the $\mu max_{(AOB)}$ was obtained as the slope of ln[NO_2^--N + NO_3^--N] over time. A Monod curve of specific COD degradation rate was constructed from the COD degradation over time by different food-microorganism ratios (So/Xo). The maximum specific COD degradation rate and $Ks_{(COD)}$ value were obtained from this Monod curve. The $\mu max_{(HET)}$ was computed as the product of the maximum specific COD degradation rate and Y_H [13]. The Y_H was calculated as the ratio of the VSS produced over COD consumed. Finally, the b_H was obtained following the Ezechi et al. [16] study by the exponential decay method. The slope of the $\ln[VSS_{(t)} - VSS_{(u)}]$ over time is the b_H value. $VSS_{(t)}$ and $VSS_{(U)}$ are the VSS concentrations at time t and the end of the batch experiment, respectively.

2.4. Calibration and validation

The built-in ASM from the BioWin software was used to perform simulations. All acquired wastewater characteristics, operational parameters, and kinetic parameters from batch experiments were applied to the model. Under steady-state condition, the model was further calibrated by fine-tuning other stoichiometric and kinetic parameters to fit the average effluent concentrations. After the steady-state simulation results agreed with the measured effluent values, the dynamic simulation was performed [17]. The model output was confirmed in dynamic simulation over time-varying flow and loading. The model's goodness-of-fit to the measured effluent concentration was estimated using Theil inequality coefficient (TIC), see Equations (2). A value of TIC ≤ 0.3 is deemed acceptable [2].

$$TIC = \frac{\sqrt{\sum_{i=1}^{N}(m_i - p_i)^2}}{\sqrt{\sum_{i=1}^{N} m_i^2} + \sqrt{\sum_{i=1}^{N} p_i^2}} \quad (2)$$

where mi and pi are the measured and simulated values for specific water quality, respectively, and N is the number of data points.

2.5. Optimization

The validated model was then used to create a modified configuration where the overall anoxic hydraulic retention time increased. Then, an optimization study was done with the aid of response surface methodology (RSM). RSM analyzed the effects of factors on chosen responses. A central composite design was adopted for the experimental design. The factors were the operating parameters, and the responses were TN removal efficiency and operating cost. The operating cost comprised the methanol dosage cost and electricity cost from mixers and blowers.

$$\text{Operating cost [NT\$/m}^3] = \frac{y\, C_{MC}}{V_T} + \frac{24\, z\, C_{EC}}{V_T} \quad (3)$$

where y and z are the prices of the methanol and electricity, C_{MC} is the methanol flowrate (L/d), V_T is the treated wastewater (m^3/d), and C_{EC} is the power from blowers and mixers (kW). The price of 45% methanol was 7.1262 NT$/L, and the average electricity price was 3.1154 NT$/kWh [18]. The mixer's power per unit volume was 14.6 W/m^3, and the blower's power was computed according to BioWin's adiabatic equation. The software itself computed the overall power.

A second-order polynomial regression model described the relationship between factors and a response. The regression model was validated with the report of ANOVA, coefficient of determination (r^2), adjusted r^2, and predicted r^2. The operating parameters were optimized simultaneously using Minitab software to maximize the TN removal efficiency and minimize the operating cost.

3. Results and discussion

3.1. Wastewater characterization

Table 2 summarizes the average results from a three-month sampling period. The pH, total chemical oxygen demand (TCOD), suspended solids (SS), ammonia-nitrogen (NH_4^+-N), nitrite-nitrogen (NO_2^--N), and total nitrogen (TN) fall under the standard discharge limit.

Table 2. Wastewater characteristics.

Wastewater parameters	Units	Average influent values	Average effluent values	Standard discharge limit [19]
pH	-	7.5 ± 0.1	7.0 ± 0.1	6.0-9.0
TCOD	mg-O_2/L	97.6 ± 6.6	24.1 ± 3.5	100
SS	mg-SS/L	36.4 ± 6.7	0.8 ± 1.0	30
VSS	mg-VSS/L	30.1 ± 6.4	0.5 ± 0.9	-
NH_4^+-N	mg-N/L	36.0 ± 3.7	0.7 ± 0.5	6
NO_2^--N	mg-N/L	0.0 ± 0.0	0.6 ± 0.4	50
NO_3^--N	mg-N/L	0.6 ± 0.3	7.2 ± 1.6	-
TKN	mg-N/L	47.0 ± 7.2	4.1 ± 0.9	-
TN	mg-N/L	47.6 ± 7.2	11.8 ± 2.1	20
TP	mg-P/L	3.8 ± 0.9	2.9 ± 0.7	-

YK WWTP has a very low influent TCOD/TN with an average of 2.0, indicating limiting denitrification performance in the BNR system. To enhance the denitrification performance, the YK WWTP incorporates 45% methanol as an external carbon source. With an average methanol flow of 1,970.8 L/d, the TN final concentration reached an average of 11.8 mg/L. Adding methanol inevitably increased the operating costs – a supplementary average of 14,044 NT$/d. Therefore, this study explored the cost reduction for the YK BNR process.

The influent COD fraction was utilized in the calibration of the model. The average readily biodegradable COD, soluble non-biodegradable COD, slowly biodegradable COD, and particulate non-biodegradable COD were 0.25 ± 0.04, 0.24 ± 0.02, 0.12 ± 0.06, and 0.39 ± 0.08, respectively.

3.2. Batch experiment for ASM parameter

In the nitrification batch experiment, the maximum specific growth rate for ammonia-oxidizing bacteria (AOB) ($\mu max_{(AOB)}$) was determined (Figure 2). NO_2^--N and NO_3^--N productions indicate AOB's activity [15]. The linear regression slope was 3.37 d^{-1}, representing $\mu max_{(AOB)}$.

In carbon addition batch experiments, the maximum specific growth rate of heterotrophic bacteria ($\mu max_{(HET)}$), substrate (COD) half-saturation ($Ks_{(COD)}$), and growth yield of heterotrophic under aerobic state (Y_H) were obtained. A monod curve of specific COD degradation rate was constructed from the COD degradation over time by different food-microorganism ratios (So/Xo). The MATLAB curve fitting tool was utilized to fit the monod curve, and the result is shown in Figure 3.a. The maximum specific COD degradation rate and $Ks_{(COD)}$ were 4.6442 mg-COD/mg-VSS-day and 0.7706 mg-COD/L, respectively. The Y_H was 0.33 mg-VSS/mg-COD, computed as the ratio of the VSS produced over COD consumed. With these outcomes, the $\mu max_{(HET)}$ can be calculated as the product of the maximum specific COD degradation rate and Y_H, resulting in a value of 1.53 d^{-1}.

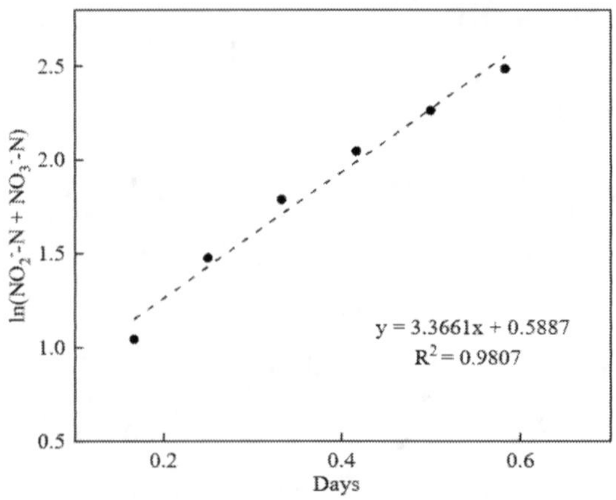

Fig. 2. Maximum specific growth rate of ammonia-oxidizing bacteria using linear regression.

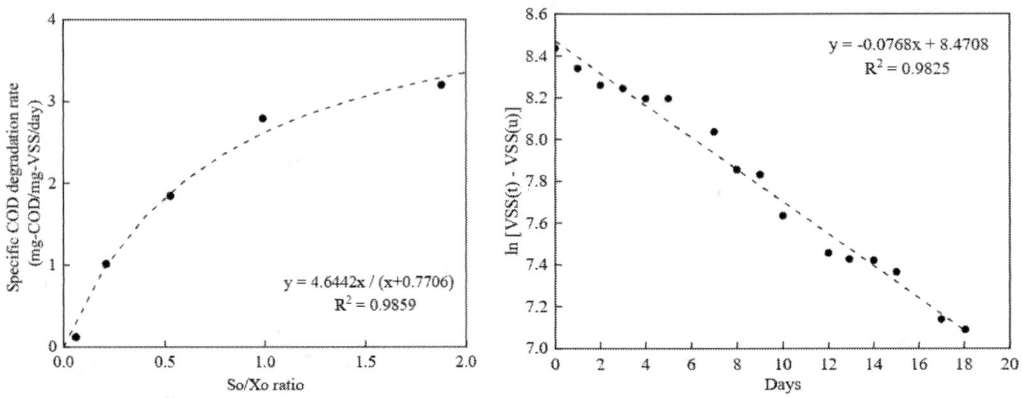

Fig. 3. (**a**) Specific COD degradation rate at different So/Xo ratios using Monod curve fitting; (**b**) Aerobic decay rate of heterotrophic bacteria using linear regression.

Finally, in the decay batch experiment, the decline of MLVSS over time is shown in Figure 3.b. Specifically, the slope of the ln[VSS(t) – VSS(u)] versus time was 0.0768 d^{-1}, representing the decay rate (b_H) under aerobic condition. The value agrees with a proposed estimation of b_H by Keppeler and Gujer [20], which can be computed as 5% of $umax_{(HET)}$.

3.3. Steady-state simulation

To begin with the simulation, all acquired information of YK BNR, including influent wastewater characteristics; BNR configuration, design, and operating parameters; and biokinetics from batch experiments, were constructed and applied in the BioWin software.

Under steady state condition, other stoichiometric and kinetic parameters were adjusted to calibrate the BioWin model – fitting the model effluent concentration to the average measured effluent. The adjusted parameters were those that exhibited a major effect on the effluent concentration of the YK BNR process.

For NH_4^+-N concentration, the substate half-saturation constant and the aerobic decay rate of ammonia-oxidizing bacteria were adjusted. The substate half saturation constant was changed to 10, which is close to the adjusted value in the study by Satoh et al. [21]. Meanwhile, the decay rate was changed to 0.55, which agrees with Elawwad et al. [10], who increased it to 0.33, and Hu et al. [22], who increased it to 0.75.

As for the calibration of NO_2^--N and TKN concentrations, the substrate half saturation of nitrite-oxidizing bacteria and the ammonification rate were adjusted, respectively. The former parameter was increased to 1.36. This agrees with Elawwad et al. [23], who changed it to 1.5. The latter parameter decreased to 0.03, close to the changes made by Bashar [24] at 0.04.

As for NO_3^--N concentration, the DO half saturation of heterotrophic bacteria, growth yield under anoxic state of heterotrophic bacteria, anoxic growth factor of heterotrophic bacteria, and anoxic hydrolysis factor were adjusted. The DO half saturation was increased to 0.4, close to the Sager et al. [25] study. The yield (aerobic) value was referred from 0.33 mg-VSS/mg-COD of batch experiment result and converted to 0.4686 mg-COD/mg-COD, assuming 1 mg-VSS = 1.42 mg-COD. The yield (anoxic) was decreased with respect to the yield (aerobic), given that the anoxic yield is usually lower than the aerobic one [26]. Anoxic growth factor and anoxic hydrolysis factors are within the ranges reported by Orhon et al. [27] and Hulsbeek et al. [28], respectively.

Since methanol was being added as an external carbon source to the BNR of YK, the kinetic and stoichiometry of methylotrophic bacteria were also explored to calibrate the NO_3^--N concentration. The use of methanol under anoxic conditions in the BioWin model is limited to methylotrophic bacteria [8]. Once the appropriate population of the said bacteria is established, the methanol half saturation constant in typical municipal WWTP is usually very low [8]. Hence, it was decreased to 0.13 mg/L, which is within the range of 0.03 – 2.35 mg/L [29]. The anoxic yield of the corresponding bacteria was the last parameter to be lowered to finally match the simulated NO_3^--N concentration with the measured effluent concentration.

Tables 3 and 4 show the summarized calibrated values and results under steady state, respectively. All wastewater quality parameters are less than 20% relative error, indicating a good agreement between the simulated result and the measured one.

Table 3. Calibrated stoichiometric and kinetics parameters.

Parameters	BioWin default value	Batch experiment results and adjusted values	Remarks
Ammonia oxidizing bacteria			
Maximum specific growth rate [1/d]	0.90	3.3661	Experiment
Substrate (NH_4^+-N) half-saturation [mg-N/L]	0.70	10.0	Adjusted
Aerobic decay rate [1/d]	0.17	0.55	Adjusted
Nitrite oxidizing bacteria			
Substrate (NO_2^--N) half-saturation [mg-N/L]	0.10	1.36	Adjusted
Ordinary heterotrophic bacteria			
Maximum specific growth rate [1/d]	3.20	1.5326	Experiment
Substrate (COD) half-saturation [mg-COD/L]	5.00	0.7706	Experiment
Anoxic growth factor [-]	0.5	0.94	Adjusted
Aerobic decay rate [1/d]	0.62	0.0768	Experiment
DO half-saturation [mg/L]	0.15	0.4	Adjusted
Growth yield (aerobic) [mg-COD/mg-COD]	0.666	0.4686	Experiment
Growth yield (anoxic) [mg-COD/mg-COD]	0.54	0.21	Adjusted
Methylotrophic bacteria			
Methanol half-saturation [mg-COD/L]	0.5	0.13	Adjusted
Growth yield (anoxic) [mg-COD/mg-COD]	0.4	0.20	Adjusted
Common bacteria			
Anoxic hydrolysis factor [-]	0.28	1.0	Adjusted
Ammonification rate [L/(mg-COD-d)]	0.08	0.03	Adjusted

Table 4. Steady State Simulation.

Parameters	Average measured value	Model result	Relative Error (%)
TCOD	24.13	23.69	1.82
NH_4^+-N	0.72	0.72	0.00
NO_2^--N	0.60	0.70	16.67
NO_3^--N	7.16	7.34	2.51
TKN	4.07	4.14	1.72
TN	11.83	12.19	3.04
TP	2.92	3.15	7.88

3.4. Dynamic state simulation

Dynamic simulation was performed over time-varying flow and loading. The data on the influent and effluent of BNR were collected from March to May of 2023. Figure 4 shows the dynamic simulation results for TCOD, TKN, TN, and TP. All Theil inequality coefficients (TIC) were lower than 0.3 of the acceptable limit. Hence, the model and the measured values for carbon, nitrogen, and phosphorus compounds are in good agreement, and the model can be used for optimization.

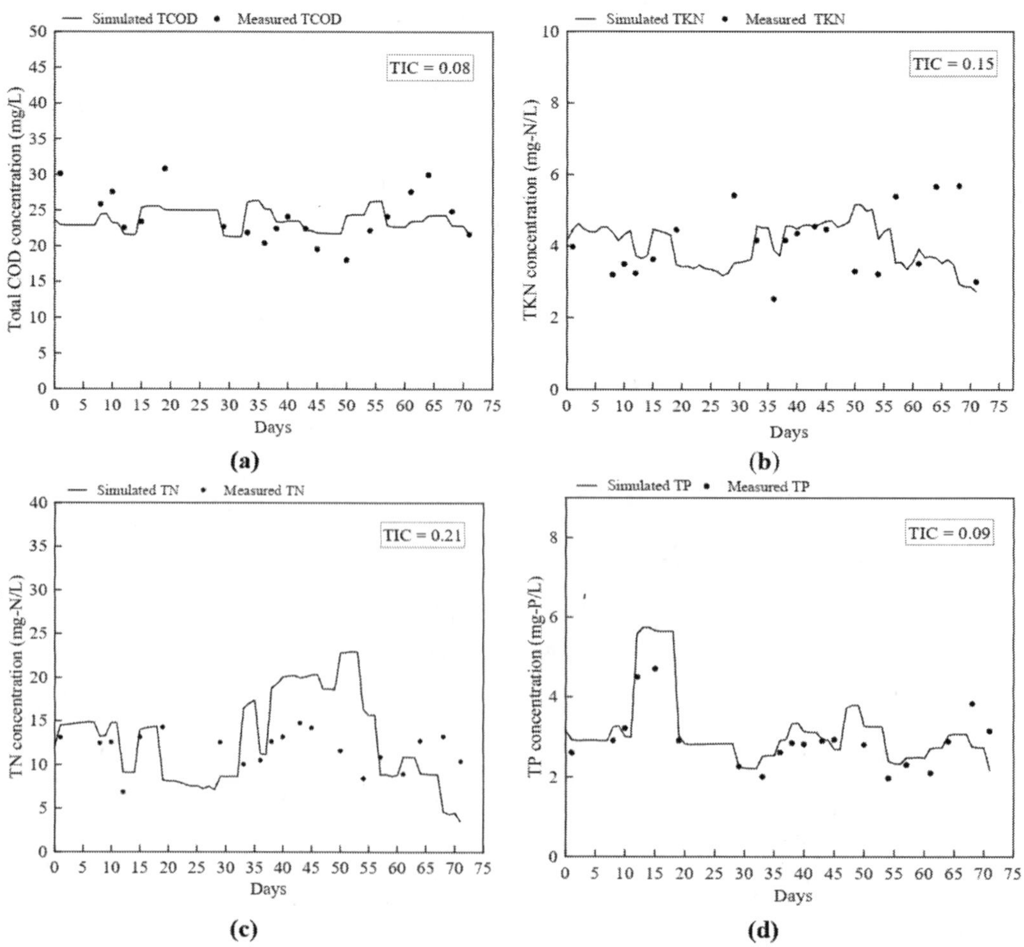

Fig. 4. Measured versus dynamic simulated values of **(a)** TCOD, **(b)** TKN, **(c)** TN, **(d)** TP.

3.5. Optimization

This study evaluated and optimized the operating parameters of YK BNR to reduce operational costs while improving TN removal. Since the removal efficiency for NH_4^+-N was already high (averaging 97.9%), and the TN of the effluent consisted of 61% nitrate-nitrogen, the focus for TN removal was through the denitrification process. The operating parameters that greatly affect denitrification are dissolved oxygen (DO), solid retention time (SRT), and carbon amount (in this case, methanol flowrate) [1,5,30]. The methanol distribution ratio between the two methanol tanks was also included. This was to account for the amount of carbon source distributed in the 2nd and 6th biological tanks. While the chosen operating parameters were evaluated, the air flowrate in the MBR was kept constant, as this rate is influenced by the need for membrane scouring to reduce fouling. Also, the influent ratio distribution remained the same at 2:3:1.

Another way of optimizing the denitrification process is to increase the anoxic hydraulic retention time (HRT). HRT determines the contact time between the microorganism and organic substrate to biodegrade a pollutant simultaneously [30]. For denitrification, HRT is directly related to nitrate removal [31]. A higher anoxic HRT gives more microbial reaction time for the denitrifiers to convert nitrate to nitrogen gas. With this information, a modified configuration of the YK BNR was explored with minimal changes. The 3rd and 4th biological tanks were replaced by oxic and anoxic conditions, respectively (Figure 5). By replacing the 4th tank with anoxic, the overall anoxic tank volume increased by 20.8%, thereby increasing the anoxic HRT by 20.9%.

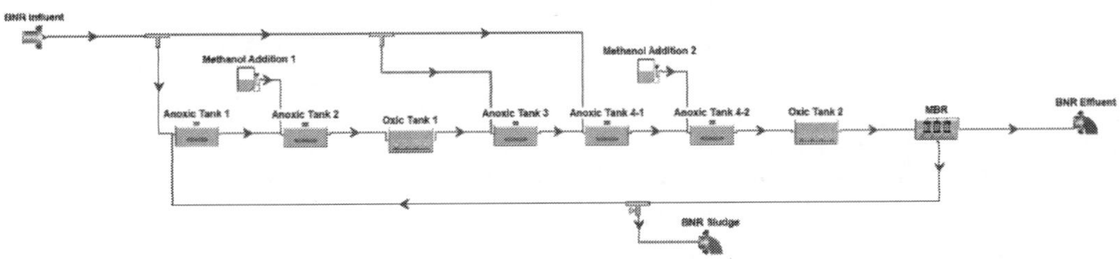

Fig. 5. Modified BNR configuration for YK municipal WWTP (exported from BioWin software).

Using response surface methodology (RSM), the effects of factors (i.e., operating parameters) on the responses (i.e., removal efficiency of TN and operating cost) can be easily observed. With a central composite design full factorial, 52 different operational scenarios were created according to five factors. The corresponding range values of the five factors are summarized in Table 5. The BioWin calibrated model ran each scenario, and the results are shown in Table A1. The second-order polynomial equations, which describe the relationship between the factors and responses, are shown in Table 6.

Table 5. Operating parameters and their ranges for RSM

Factors for RSM (uncoded parameter)	Coded parameter	Modified configuration range values
DO of Oxic Tank 1 [mg/L]	A	0.5 – 2.0
DO of Oxic Tank 2 [mg/L]	B	0.5 – 2.0
SRT [days]	C	15 - 45
Methanol distribution ratio [-]	D	0:1 – 1:1 [0 – 1] [a]
Methanol flowrate [L/d]	E	1000 - 1970.8

[a] For the methanol ratio, 0 – 1 was utilized as inputs in RSM representing 0:1 – 1:1.

Table 6. Polynomial equations for the responses in modified configuration RSM design.

Response	Polynomial equation [a]
TN %RE	$30.26 - 4.26\,A - 3.75\,B + 0.309\,C - 1.17\,D + 0.02641\,E + 1.29\,A^2 + 1.54\,B^2 + 0.00067\,C^2 + 0.81\,D^2 + 0.000001\,E^2 + 0.292\,A{*}B - 0.0643\,A{*}C + 1.001\,A{*}D + 0.000379\,A{*}E + 0.0254\,B{*}C + 1.200\,B{*}D - 0.003403\,B{*}E - 0.0492\,C{*}D - 0.000130\,C{*}E - 0.001401\,D{*}E$
Operating Cost [NT$/m³]	$0.94178 + 0.01683\,A + 0.03960\,B - 0.000840\,C + 0.00002\,D + 0.000384\,E + 0.00040\,A^2 - 0.00082\,B^2 + 0.000006\,C^2 - 0.00007\,D^2 - 2.69\times10^{-10}\,E^2 - 0.001823\,A{*}B + 0.000092\,A{*}C - 0.001309\,A{*}D - 0.000003\,A{*}E - 0.000022\,B{*}C - 0.000493\,B{*}D + 0.000002\,B{*}E + 0.000086\,C{*}D + 7.7\times10^{-8}\,C{*}E + 2.8\times10^{-7}\,D{*}E$

[a] A – DO of oxic tank 1; B – DO of oxic tank 2; C – SRT; D – Methanol ratio; E – Methanol flowrate.

The TN and operating cost regression models fit well with the data (Table 7). The coefficient of determination (R^2), adjusted R^2, and predicted R^2 were either equal to 1 or close to 1. The high value of predicted R^2 means great predictive ability, which can help predict the optimized parameters. The statistical analysis was performed at a confidence level of 95% ($p < 0.05$). The p-values of the models were all <0.0001, indicating that the models were significant and could be used to navigate the design space.

Table 7. ANOVA regression model for TN and operating cost for modified configuration.

Regression model	F-value	P-value	R^2	R^2 (adjusted)	R^2 (predicted)
TN	165.55	< 0.0001	0.9907	0.9847	0.9647
Operating Cost	38359.39	< 0.0001	1.0000	0.9999	0.9998

In the TN removal efficiency response model, the p-values of all linear terms (i.e., A, B, C, D, E) and interaction terms of AC, BD, BE, and CE were all significant ($p < 0.05$). The rest of the terms, especially all square terms, were insignificant. From the F-values of linear terms, the order of significance that affected TN removal efficiency was as follows: E > B > C > A > D. Methanol flowrate played a major role compared to other terms. In denitrification, the carbon source serves as the electron donor for nitrate reduction. It is an essential substance that directly impacts the rate and effect of denitrification [31]. Hence, the methanol flow rate, a supplemental carbon in YK BNR WWTP, provides sufficient electron donors to enhance the denitrification process, thereby having the greatest potential to lower the TN in the effluent.

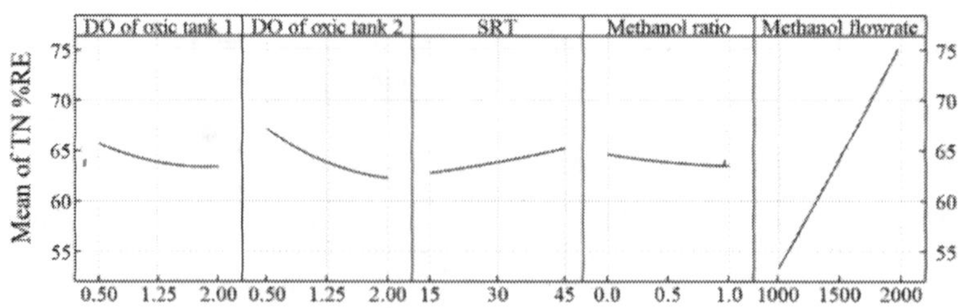

Fig. 6. Main effects plot for TN %RE (exported from Minitab software).

Furthermore, the main effects plot shows the trend and relative strength of the factors to the response. In line with the statistical result, the methanol flowrate had the greatest effect, with approximately a 22% change in the TN removal efficiency from 1,000 L/d to 1,970.8 L/d (Figure 6). In contrast, the methanol ratio had the lowest effect. As the figure shows, the TN removal efficiency was directly proportional to the methanol flowrate and indirectly proportional to the methanol ratio. The methanol ratio 0:1 (i.e., all flow directed to anoxic tank 4-2) was more favorable in lowering TN concentration. The nitrate forms from oxic tank 1 needed more external carbon for denitrification to succeed in subsequent anoxic tanks. As for both DO from oxic tanks 1 and 2, the graph has a curve where lower DO was more favorable to denitrification. This agrees with the literature where a decrease in DO increases the denitrification rate [31]. Also, a reduction of DO in an aerobic tank can make the reflux of DO to an anoxic tank decrease as well, making denitrification favorable [2]. Lastly, a higher SRT was more favorable in denitrification. This might be due to a higher growth rate of denitrifying bacteria at 45 days of SRT.

In the operating cost response model, the p-values of all linear terms (i.e., A, B, C, E) except D and interaction terms of AB, AC, AD, AE, BE, CD, and CE were all significant ($p < 0.05$). Similar to the TN removal efficiency model, all square terms were insignificant. From the F-values of linear terms, the order of significance was as follows: E > B > A > C. Methanol flowrate had a huge influence on the operating cost compared to other factors, which was further proven in the main effects plot (Figure 7). The operating cost was directly proportional to the methanol flowrate and DO of both oxic tanks. Higher utilization of methanol and higher air flowrate incur additional costs. A higher air flowrate is needed to increase the DO concentration and vice versa.

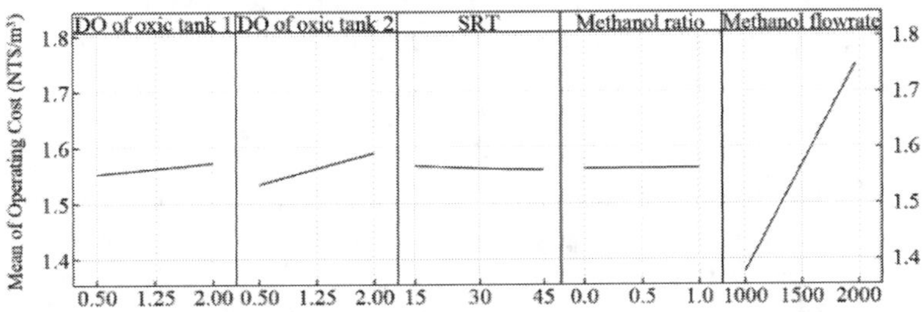

Fig. 7. Main effects plot for operating cost (NT$/m^3) (exported from Minitab software).

The response surface plots derived from the regression equations show the relationship between the two interacting factors and their effect on response. Figure 8 shows the interaction between the top two significant factors (i.e., DO of oxic tank 2 and methanol flowrate). The remaining factors not shown in a single plot kept their settings at their middle value. The maximum TN removal efficiency occurred at the intersection of the minimum DO of oxic tank 2 and maximum methanol flowrate (Figure 8.a.). A low operating cost occurred at the intersection between lower values of DO of oxic tank 2 and methanol flowrate (Figure 8.b).

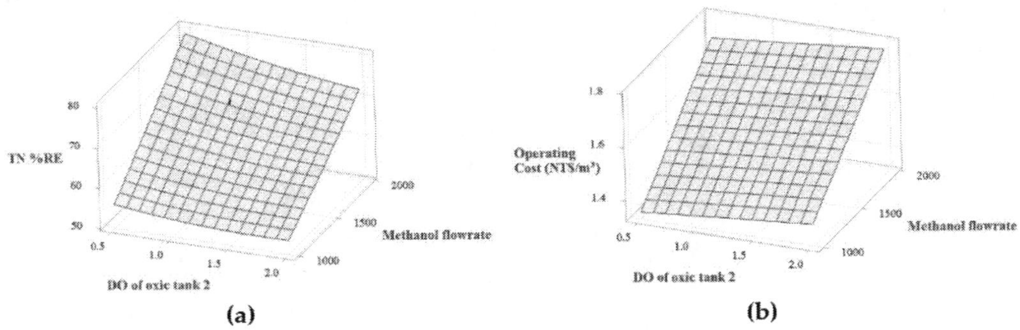

Fig. 8. Response surface plot of (**a**) TN%RE (B vs E); (**b**) Operating cost (B vs E) (exported from Minitab software)

Using the response optimizer function of the Minitab software where TN removal efficiency was maximized and operating cost was minimized, the optimum operating parameters for the modified BNR configuration were as follows: 0.5 mg/L DO for oxic tanks, 45 days of SRT, 0:1 methanol distribution ratio, and 1,794.29 L/d methanol. Under these operating parameters, the BioWin calibrated model resulted in 86.31% TN removal efficiency and 1.63 NT$/m^3 operating cost. The TN removal efficiency increased by 16% from 74.4%RE and the operating cost decreased by 4.12%. The methanol flowrate was reduced by 8.96%. Even though the NH_4^+-N removal efficiency decreased slightly to 96.94%, the corresponding concentration of 1.1 mg/L NH_4^+-N at the effluent was still far below the standard discharge limit.

4. Conclusion

The study simulated the step-feed AO-MBR BNR process with the methanol addition of YK WWTP. The methanol, utilized as the external carbon source, was costly and had been the concern of the YK WWTP. This study lessened its use and lowered the operating cost while still improving the TN removal efficiency. The BNR process model was created using BioWin's built-in ASM. This model was calibrated using the results of biokinetic batch experiments and numerical adjustments of stoichiometric and kinetic parameters to imitate the performance of the BNR process. The model was validated using dynamic simulation and was in good agreement with the measured effluent values according to the Theil inequality coefficient. Using the calibrated model, a modified configuration was proposed. The 3rd and 4th biological tanks were replaced by oxic and anoxic conditions, respectively. This replacement increased the anoxic tank volume by 20.8%, thereby increasing the anoxic HRT by 20.9%. The modification resulted in an improved denitrification process. With the aid of RSM, the operating parameters of the modified configuration were optimized to achieve maximum TN removal efficiency and minimum operating cost. The optimum operating parameters were 0.5 mg/L DO concentration, 45 days of SRT, 0:1 methanol distribution ratio, and 1,794.29 L/d methanol flowrate. All contributed to the improvement of the denitrification process. With the optimum operating parameters, the TN removal efficiency increased by 16% (86.31 %RE), and the operating cost decreased by 4.12% (1.63 NT$/m3). The methanol usage was reduced by 8.96%.

Appendix

Table A1. Experimental design results for TN removal efficiency and operating cost

Run	DO of oxic tank 1 [mg/L]	DO of oxic tank 2 [mg/L]	SRT [day]	Methanol ratio	Methanol Flowrate [L/d]	TN %RE	Operating Cost [NT$/m^3]
1	0.50	0.50	15	0.00	1000.00	55.21	1.34
2	2.00	0.50	15	0.00	1000.00	53.19	1.37
3	0.50	2.00	15	0.00	1000.00	52.31	1.40
4	2.00	2.00	15	0.00	1000.00	51.11	1.42
5	0.50	0.50	45	0.00	1000.00	61.70	1.33
6	2.00	0.50	45	0.00	1000.00	56.87	1.36
7	0.50	2.00	45	0.00	1000.00	58.57	1.39
8	2.00	2.00	45	0.00	1000.00	54.83	1.41
9	0.50	0.50	15	1.00	1000.00	55.31	1.34
10	2.00	0.50	15	1.00	1000.00	53.32	1.37
11	0.50	2.00	15	1.00	1000.00	52.37	1.40
12	2.00	2.00	15	1.00	1000.00	51.26	1.42
13	0.50	0.50	45	1.00	1000.00	59.76	1.33
14	2.00	0.50	45	1.00	1000.00	56.32	1.36
15	0.50	2.00	45	1.00	1000.00	56.72	1.39
16	2.00	2.00	45	1.00	1000.00	54.24	1.41
17	0.50	0.50	15	0.00	1970.80	81.56	1.72
18	2.00	0.50	15	0.00	1970.80	83.56	1.73
19	0.50	2.00	15	0.00	1970.80	73.12	1.77
20	2.00	2.00	15	0.00	1970.80	71.21	1.79

Table A1. Experimental design results for TN removal efficiency and operating cost (cont.)

Run	DO of oxic tank 1 [mg/L]	DO of oxic tank 2 [mg/L]	SRT [day]	Methanol ratio	Methanol Flowrate [L/d]	TN %RE	Operating Cost [NT$/m³]
21	0.50	0.50	45	0.00	1970.80	87.74	1.70
22	2.00	0.50	45	0.00	1970.80	77.93	1.73
23	0.50	2.00	45	0.00	1970.80	76.21	1.77
24	2.00	2.00	45	0.00	1970.80	73.10	1.78
25	0.50	0.50	15	1.00	1970.80	79.55	1.72
26	2.00	0.50	15	1.00	1970.80	79.90	1.73
27	0.50	2.00	15	1.00	1970.80	73.35	1.77
28	2.00	2.00	15	1.00	1970.80	72.20	1.79
29	0.50	0.50	45	1.00	1970.80	78.98	1.71
30	2.00	0.50	45	1.00	1970.80	77.49	1.73
31	0.50	2.00	45	1.00	1970.80	74.44	1.77
32	2.00	2.00	45	1.00	1970.80	73.16	1.78
33	0.50	1.25	30	0.50	1485.40	65.25	1.55
34	2.00	1.25	30	0.50	1485.40	63.57	1.57
35	1.25	0.50	30	0.50	1485.40	65.75	1.54
36	1.25	2.00	30	0.50	1485.40	63.33	1.59
37	1.25	1.25	15	0.50	1485.40	62.60	1.57
38	1.25	1.25	45	0.50	1485.40	65.06	1.56
39	1.25	1.25	30	0.00	1485.40	63.86	1.56
40	1.25	1.25	30	1.00	1485.40	63.90	1.56
41	1.25	1.25	30	0.50	1000.00	54.05	1.38
42	1.25	1.25	30	0.50	1970.80	74.00	1.75
43	1.25	1.25	30	0.50	1485.40	63.96	1.56
44	1.25	1.25	30	0.50	1485.40	63.96	1.56
45	1.25	1.25	30	0.50	1485.40	63.96	1.56
46	1.25	1.25	30	0.50	1485.40	63.96	1.56
47	1.25	1.25	30	0.50	1485.40	63.96	1.56
48	1.25	1.25	30	0.50	1485.40	63.96	1.56
49	1.25	1.25	30	0.50	1485.40	63.96	1.56
50	1.25	1.25	30	0.50	1485.40	63.96	1.56
51	1.25	1.25	30	0.50	1485.40	63.96	1.56
52	1.25	1.25	30	0.50	1485.40	63.96	1.56

Acknowledgments

The authors would like to acknowledge the financial support from the National Science and Technology Council of Taiwan [NSTC 112-2221-E-006-034-MY3] and National Land Management Agency of Ministry of Interior. The authors would like to acknowledge the technical support and information sharing from Forest Water Environmental Engineering Co. Ltd. A.C.T. extends her gratitude for the financial support from the Department of Science and Technology–Science Education Institute (DOST-SEI) and Engineering Research and Development for Technology (ERDT) of the Philippines.

References

1. X. Mao, P. H. Myavagh, S. Lotfikatouli, B. S. Hsiao, and H. W. Walker, Journal of Environmental Engineering **146**, (2020)
2. J. Cao, E. Yang, C. Xu, T. Zhang, R. Xu, B. Fu, Q. Feng, F. Fang, and J. Luo, Science of the Total Environment **769**, 144851 (2021)
3. A. Elawwad, M. Zaghloul, and H. Abdel-Halim, Journal of Water Reuse and Desalination **7**, 37 (2017)
4. K. V. Gernaey, M. C. M. Van Loosdrecht, M. Henze, M. Lind, and S. B. Jørgensen, Environmental Modelling and Software **19**, 763 (2004)
5. Z. Wang, T. Ji, Z. Zhu, P. Mei, W. Guo, and Y. Li, Water (Switzerland) **14**, (2022)
6. H. Dai, W. Chen, Z. Dai, X. Li, and X. Lu, Environmental Science and Pollution Research **24**, 19211 (2017)

7. S. Huang, C. K. Pooi, X. Shi, S. Varjani, and H. Y. Ng, Science of the Total Environment **747**, 141311 (2020)
8. EnviroSim Associates Ltd, (2017). BioWin Help Manual. Available online: https://envirosim.com/4685 (accessed on 1 March 2023)
9. E. Liwarska-Bizukojc and R. Biernacki, Bioresour Technol **101**, 7278 (2010)
10. A. Elawwad, M. Matta, M. Abo-Zaid, and H. Abdel-Halim, Journal of Water Process Engineering **31**, 100819 (2019)
11. S. Gillot, T. Ohtsuki, L. Rieger, A. Shaw, I. Takács, and S. Winkler, Influents **4**, (2009)
12. R. B. Baird, A. D. Eaton, E. W. Rice, and L. Bridgewater, *Standard Methods for the Examination of Water and Wastewater*, 23rd Edition (Washington, D.C. : American Public Health Association, 2017)
13. G. A. Ekama, P. L. Dold, and G. R. Marais v., Water Science and Technology **18**, 91 (1986)
14. D. Mamais, D. Jenkins, and P. Prrr, Water Res **27**, 195 (1993)
15. P. Antoniou, J. Hamilton, B. Koopman, R. Jain, B. Holloway, G. Lyberatos, and S. A. Svoronos, Water Res **24**, 97 (1990)
16. E. H. Ezechi, S. R. B. M. Kutty, M. H. Isa, A. Malakahmad, C. M. Ude, E. J. Menyechi, and E. Olisa, Res J Microbiol **10**, 355 (2015)
17. P. A. Vanrolleghem, G. Insel, B. Petersen, G. Sin, D. De Pauw, I. Nopens, H. Dovermann, S. Weijers, and K. Gernaey, Proceedings of the Water Environment Federation **2003**, 210 (2012)
18. FOCUS TAIWAN, 2023 (2023). Available online: https://focustaiwan.tw/business/202303170024 (accessed on 1 May 2023).
19. Laws & Regulations Database of The Republic of China (Taiwan) Table 14 (2019). Available online: https://law.moj.gov.tw/ENG/LawClass/LawAll.aspx?pcode=O0040004 (accessed on 1 March 2023).
20. J. Kappeler and W. Gujer, Water Science and Technology **25**, 125 (1992)
21. H. Satoh, E. Okuda, T. Mino, and T. Matsuo, Water Science and Technology **42**, 29 (2000)
22. X. Hu, L. Xie, C. Mi, and D.H. Yang, Journal of Zhejiang University: Science A **15**, 743 (2014)
23. A. Elawwad, M. Abo-Zaid, and M. Edward, in *Proceedings of the International Conference of Recent Trends in Environmental Science and Engineering (RTESE'17)* (2017), p. 139
24. R. Bashar, Evaluation of Phosphorus Removal/Recovery Processes during Municipal Wastewater Treatment, PhD Dissertation, UNIVERSITY OF WISCONSIN-MADISON, 2018
25. A. Sager, L. Knapp, S. Ergas, and G. Iranipour, Florida Water Resources Journal 22 (2016)
26. P. S. Barker and P. L. Dold, Water Environment Research **69**, 969 (1997)
27. D. Orhon, S. Sozen, and N. Artan, Water Science and Technology **34**, 67 (1996)
28. J. J. W. Hulsbeek, J. Kruit, P. J. Roeleveld, and M. C. M. Van Loosdrecht, Water Science and Technology **45**, 127 (2002)
29. V. Torres, I. Takács, R. Riffat, A. Shaw, and S. Murthy, in *Proceedings of the Water Environment Federation* (2012), pp. 3234–3244
30. Z. Z. Loh, N. S. Zaidi, A. Syafiuddin, E. L. Yong, M. B. Bahrodin, A. Aris, and R. Boopathy, Bioresour Technol Rep **23**, 101505 (2023)
31. X. Fu, R. Hou, P. Yang, S. Qian, Z. Feng, Z. Chen, F. Wang, R. Yuan, H. Chen, and B. Zhou, Science of the Total Environment **817**, 153061 (2022)

Soil Moisture Content Sensor: Utilizing Bacterial Cells as Living Sensors

Trang Nakamoto[1] *, *Dung Nakamoto*[1], *Kenji Tamenori*[1], *and Kozo Taguchi*[1]
[1]Department of Electrical and Electronic Engineering, Ritsumeikan University, 1-1-1 Noji-higashi, Kusatsu, Shiga, Japan

Abstract. Soil moisture affects water and heat exchange between soil and air, weather and climate, and plant growth. Therefore, controlling and predicting soil moisture is vital for irrigation, nutrient supply, absorption, and crop productivity. Soil microbial fuel cell (SMFC) is a recent technology that generates electricity from soil microorganisms and chemicals. SMFC can be used to sense soil moisture and treat polluted soil. In this study, we designed a low-cost, portable, and easy-to-install SMFC for soil moisture sensing, and it was tested at four moisture levels: 40%, 60%, 80%, and 100% soil water holding capacity (SWHC). The SMFC worked best at 60 - 80% SWHC, which is suitable moisture for many plants. The results suggest that the proposed SMFC can be a potential soil moisture sensor.

1 Introduction

Water is essential for the ecological balance and life on Earth [1]. Soil moisture affects the land ecosystem and varies in different regions due to climate and weather. Soil moisture is essential for plant growth, especially in agriculture, as it provides or controls nutrients and temperature. It influences biophysical processes, such as seed germination and plant transpiration. It also enhances the soil's biological activities, improving plant nutrition. However, excess water causes high methane emissions and hinders plant growth due to anaerobic conditions [2]. Water-saturated soil is not suitable for most crops. Therefore, optimal soil water levels are necessary for high production. Water shortage or surplus in the soil stresses plants and reduces production.

Soil moisture measurement and monitoring are essential for specific research such as hydrology, atmosphere, weather, and climate and for improving soil and crop management. Soil moisture control affects agricultural management. Therefore, soil moisture sensing helps control irrigation and nutrient application [3, 4].

Three main methods to measure soil moisture are direct, indirect, and remote. The direct methods measure the water weight in the soil by evaporation, washing, or chemical reaction. The indirect and indirect methods estimate the water content using soil properties and other variables [5]. Both methods are accurate and cheap but slow and need special equipment. Remote methods use the data from the soil surface reflection of electromagnetic radiation to the satellite to estimate large-scale soil moisture. This method makes remote-sensing

*Corresponding author: n-trang@fc.ritsumei.ac.jp

satellites popular [6]. A study showed that soil moisture profiles were mapped by ground-penetration radar, which saves cost and time. However, this method has problems in soil surveys, data display, and maintenance [7,8]. Wireless sensor network technology can provide dynamic environmental data and interactions in a landscape. However, this technology has challenges like reactivity, robustness, network lifetime, and complex system installation [9].

Soil microbial fuel cells (SMFCs) are MFCs that use soil bacteria and organic matter in the soil to generate electricity. Conventional SMFCs have low power output because of the distance between the air-cathode and the buried anode, the soil resistance, and the electrode materials [10]. These limits their practical applications due to their complex installation and field operation [11]. Previous studies tried to solve these problems and increase power output. Different SMFC configurations, such as single-chamber and dual-chamber, have been tested with PEM.

The biofilm on the cathode and the electrode degradation affect the durability and stability of SMFCs and SMFCs-based sensors [12]. Most tested soils, such as sediment or muddy soil, are water-saturated because water lowers the resistance and improves the ionic mobility between electrodes. However, water-saturated soil is rare in the actual field.

This study proposed a portable plugged-type SMFC made by a 3D printer and low-cost materials for sensing soil moisture content (SMC). Carbon-based materials were used for the cathode and anode electrodes. The main goal was to test the SMFC power output sensitivity to the soil moisture change and use that as an electrical signal for the soil moisture sensor.

2 Experiment

2.1 SMFC fabrication

The configuration of the SMFC is shown in Fig. 1. The SMFC has a size of 10×4×2 cm, and it consists of three parts: a cathode chamber, an anode chamber, and a bottom case. The SMFC case was fabricated by a 3D printer using PLA filament.

To fabricate the cathode electrode, a paste mixture of 9 mL multiwalled carbon nanotube paint (N7006L, KJ Specialty Paper), 3 mL 10 wt% Nafion solution (Wako Pure Chemical Industries), 0.5 g carbon fiber, and 0.5 g activated carbon powder. This mixture was mixed for 24 h at 700 rpm and added to the cathode chamber. The cathode electrode was dried at 40 °C for three days. The anode electrode was made from two layers of activated carbon sheet with sizes of 4×3×0.2 cm. A cellulose sponge was loaded with Lysogeny broth (LB) medium and dried at 40 °C for two days. Then, it was attached to the anode chamber and anode electrode. This configuration of the SMFC made it easy to plug into the soil and the anode in good contact with the soil. In addition, the soil and its moisture were used as the electrolyte of the SMFC.

2.2 Experimental setup

Initially, the soil samples had 23% SMC, which is equivalent to 60% soil water holding capacity (SWHC). We tested the SMFC operation and sensing on different SMCs by setting up experiments with four soil moisture levels based on SWHC. The soil moisture was 40%, 60%, 80%, and 100% SWHC, which were equivalent to 24%, 36%, 48%, and 60% SMC in the soil.

https://doi.org/10.1051/e3sconf/202451401003

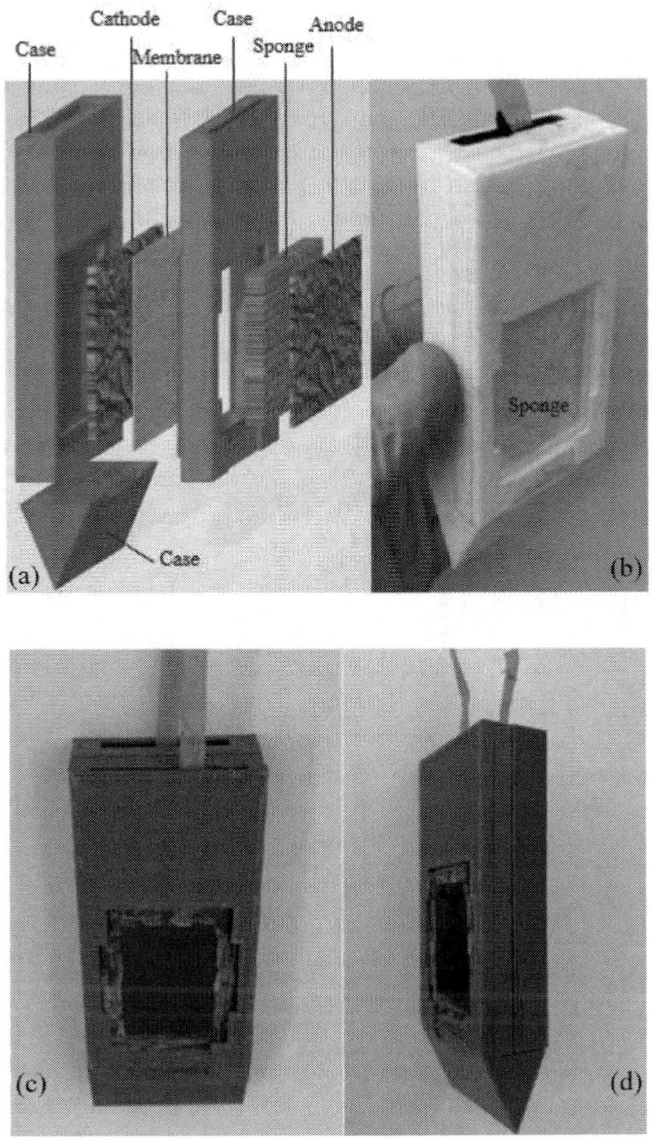

Fig. 1. The proposed SMFC configuration: schematic structure diagram of SMFC case (a) and photo images of the fabricated SMFC (b, c, d) using a 3D printer.

The experiments were done in an incubator at 30 ±2°C for 25 days. Each SMFC had a 10 kΩ external resistor connected between the two electrodes. The discharge voltage and soil moisture content changes were recorded over time. The soil water content is reduced over time because of natural evaporation. The evaporation amount was regularly measured by a scale, and water was added to compensate for the lost moisture.

All SMFCs were removed from the soil after 25 days. After that, scanning electron microscopy (SEM) was used to observe the biofilm on the anode electrodes.

3 Results and discussions

This study explores the connection between SMFC output voltage and soil moisture content. Fig. 2 shows that all four SMFCs have fast startup times, with the voltage output rising quickly in the first 7 h and then increasing further and reaching their first peaks at different times because of their different moisture level (73, 30, 15, and 105 h for SMFC1, SMFC2, SMFC3, and SMFC4, respectively). Even though no external microorganism was added, the SMFCs still started up fast. This phenomenon may be due to the LB medium-loaded sponge in the anode chamber, which gave a nutrient source to microorganisms in the soil, enhancing their activities and resulting in a short startup time. SMFC1 and SMFC2 had a similar trend after reaching their peaks at similar maximum values of 0.13 V and 0.11 V. The voltage output of SMFC1 and SMFC2 quickly dropped to 0 V and kept falling to the negative voltage side. This result may be due to the water shortage in the soil. In recent years, most SMFC research has been done in water-saturated conditions or sediment because the extra water is good for SMFC operation and stability. With 40% SWHC and 60% SWHC corresponding to 24 % - 36 % soil moisture content, this moisture level is too low to support the soil's bacteria communities and insufficient to keep the SMFC output effective. This result shows that at a level below 60 % SWHC (< 36 % SMC), the SMFC has no sensitivity to SMC at this level.

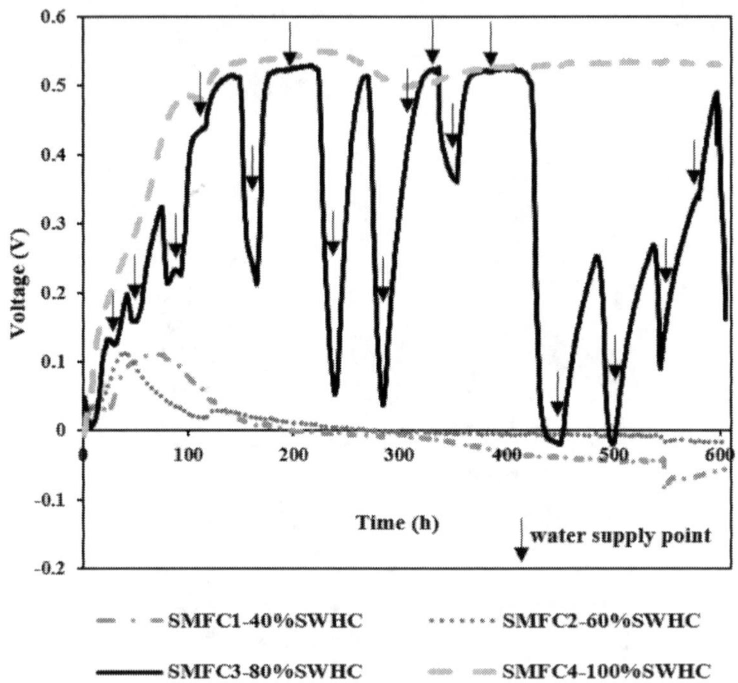

Fig. 2. Output voltages of SMFC1-4.

Unlike SMFC1 and SMFC2, the output voltage of SMFC3 showed a series of signals that changed over time, matching the change in soil moisture. Fig. 3 displays the connection between the output voltage and moisture changes during every rehydration time. At first, when the water was given daily, the SMFC3 voltage output rose slowly and reached its peak of 0.51 V after 144 h. When water adding times were about 48 h, the moisture reduced to less than 70 % SWHC, and the voltage output signal fell quickly. However, after adding water, the output could quickly return to the stable phase. The output signal went down deeper than before, especially when the water stroke was done by extending the water compensating time to 66 h. Despite this sudden drop, the output could still recover but required a longer time. Ultimately, the voltage still had the same stable value as before. This result verifies that SMFC can be a water sensor in the 60 - 80% SWHC range.

On the other hand, from the start, the electrical output of SMFC4 rose through the water supply times to the maximum value of 0.53 V and then kept at this level during the whole experiment without any noticeable changes in voltage signal because of the difference of evaporated water. Moreover, the output performance stayed stable and did not show any significant signals even after 66 h of water stroke, where the soil moisture was still at a level higher than 80% SWHC. This result may be due to excess water in the soil, which is suitable for activating and maintaining SMFC. Therefore, the result suggests that soil with high moisture at 80 - 100% SWHC (60% SMC) can make SMFC achieve a highly stable output of more than 0.5 V, but the proposed SMFC cannot sense SMC at this high SMC.

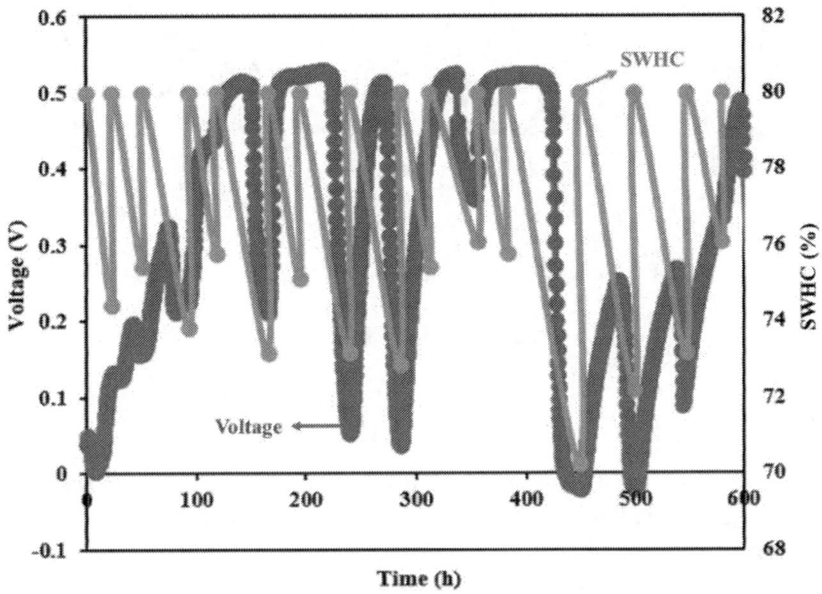

Fig. 3. The correlation between the voltage of SMFC3 and SWHC of the soil.

Crop plants research showed that low pH and high moisture content of about 80 % were suitable for disease development, while higher pH and lower moisture content of about 40 – 60 % decreased diseases [13]. Therefore, Fig. 2 and 3 results suggest that the SMFC can be an SMC sensor in the range over 60% SWHC, especially 60 – 80% SWHC corresponding to 36 - 48% SMC in the farm soil tested, in which moisture helps the plants grow and limits the diseases. Even though it cannot sense moisture at a low level, this result shows that it can improve and be a practical application in the field in the future.

The SEM images of the anode surface are shown in Fig. 4. The anode surfaces had pores with activated carbon particles clustered on carbon felts. SMFC1 and SMFC2 anodes with low soil moisture still had rod-shaped bacteria cell biofilms but in deficient amounts (Fig. 4 (a) and (b)). On the other hand, SMFC3 and SMFC4 anodes showed more biofilm (Fig. 4 (c) and (d)), which means the growth of soil microorganisms in high moisture levels is higher. The biofilm growth influenced power production, and the outputs of SMFC3 and SMFC4 were also much higher than SMFC1 and SMFC2. This result again shows the importance of water in the SMFC operation.

Fig. 4. The images of the SEM test of the anodes from SMFC1 (a), SMFC2 (b), SMFC3 (c), and SMFC4 (d) after the experiment.

4 Conclusions

We designed a low-cost, portable sensor that uses SMFC technology to measure soil moisture content. After plugging the SMFC into the soil, the output voltage will behave in a way that depends on the SMC of the soil. The sensor works best for soil moisture levels between 60 - 80% SWHC, which is equivalent to 36 - 48% SMC, and this is the optimal moisture range for many crops. The sensor is compact, portable, and sensitive to the proper soil moisture levels, so it can help farmers irrigate their crops efficiently and affordably.

References

1. M. Gavrilescu, Water **13**, (2021)
2. Y. Kamaraj, G. Punamalai, S. Kandasamy, K. Kasinathan, Arch. Micro. **202** (2020)
3. S. Unninayar, L.M. Olsen, Ref. Mod. Earth Sys. Envi. Scien. **2425** (2015)
4. P. J. Kramer, *Plant and Soil Water Relationships: A Modern Synthesis*. **45** (2015)
5. G. Kebede, S. Hunduma, Afr. J. Bas. & Appl. Scien. **12** (2020)
6. J. Acker, R. Williams, L. Chiu, P. Ardanuy, S. Miller, C. Schueler, P.W. Vachon, M. Manore, Ref. Mod. Earth Sys. Envi. Scien. **161** (2014)
7. S. Lambot, L. Weihermüller, J.A. Huisman, H. Vereecken, M. Vanclooster, E.C. Slob, Water Resour. Res. **42** (2006)
8. K. Zajícová, T. Chuman, Geoderma. **343** (2019)
9. R. C. Oliver, M. Kranz, K. Smettem, K. Mayer, Inter. J. Distri. Sen. Net. 1 (2005)
10. B. Yu, J. Tian, L. Feng, J. Haz. Mat. **336** (2017)
11. L. Hsu, A. Mohamed, P.T. Ha, J. Bloom, T. Ewing, M. Arias-Thode, B. Chadwick, H. Beyenal, J of The Electro. Soci. **164** (2017)
12. S.Z. Abbas, Y.C. Yong, F.X. Chang, Inter. J. Energy Research. **46** (2022)
13. K. Narisawa, M. Shimura, F. Usuki, S. Fukuhara, T. Hashiba, Plant Disease. **89** (2005)

Exploring Water Quality and Toxicity: An Analysis of Cadmium and Lead Concentration in the Parañaque River

Alfonso Miguel Atienza[1], and *Jan Florenz* Padilla[1], and *Nanette* Santos[1*]

[1]School of Chemical, Biological, and Materials Engineering and Sciences, Mapúa University, Intramuros, Manila, 1002, Philippines

Abstract. Rivers provide vital environmental functions, including transportation, irrigation, water supply, flood control, and habitat for aquatic life. Water contamination, resulting in poor water quality, is a pressing issue affecting communities' health, safety, and environment worldwide, leading to health issues. This research study aimed to monitor the water quality of the Parañaque River and the potential toxicity of heavy metal concentrations, specifically Cadmium and Lead, from September to October 2022. Surface water samples were taken to assess the heavy metals' concentration levels and water quality parameters: color, pH, salinity, dissolved oxygen (DO), biochemical oxygen demand (BOD), total dissolved solids (TDS), and total suspended solids (TSS). The collection of water samples followed the standard method for surface water sampling listed in the Water Quality Monitoring Manual Volume 1 of Department of Environment and Natural Resources – Environmental Management Bureau (DENR-EMB). The collected data were compared using descriptive statistics to the established water quality standards of DENR Administrative Order No. 2016-08. The analyses used a significance level of 0.05 or 5%. ANOVA and Kruskal Wallis were used to compare the sampling locations that showed no significant difference in the water quality of the three sampling sites at Parañaque River. Paired T-Test and Wilcoxon Signed Rank Test were used to compare the difference in the time of water sampling. The results showed a significant difference for all parameters between morning and afternoon sampling except DO, TSS, and TDS. The Parañaque River's water quality, heavily polluted with domestic and industrial waste, bacteria, and other microorganisms, failed to meet DENR's standards for Class C water, thus indicating its poor condition.

1 Introduction

Seventy percent of the Earth's surface is covered in water. Despite this, only 3% of the available water is freshwater, and 0.5% of the total water on Earth is available for human consumption [1]. It is accessible from different bodies of water, such as lakes, rivers, streams, and reservoirs. However, water from these sources is mainly saltwater and can only be used

* Corresponding author: ndsantos@mapua.edu.ph

after desalination [2]. Water is vital for survivability, as the human body is composed of 55% to 78% of water. It is essential in bodily and other functions such as hygiene and hydration. According to studies, population increase, growing water scarcity, industrialization, and climate change are significant issues for drinking water supply systems [3]. The rapid development of large-scale industrial operations has considerably increased human exposure to heavy metals [4]. Untreated wastewater from industries and chemical spills contains heavy metals such as copper, chromium, zinc, and manganese, which causes water pollution [5]. Due to their non-biodegradable nature and long biological half-life, the threat of heavy metals is more dangerous than other pollutants [6]. The effects of heavy metal pollution include the destruction of the aquatic environment and unsuitable water uses for drinking, agriculture, and recreation. The long-term effects of heavy metals may cause severe human health problems [7]. Quality monitoring is done to determine water contents with heavy metal contamination and its impacts on human health and aquatic life. Water quality parameters are measured to assess the water thoroughly [8]. With this, the research identifies the overall condition of the Parañaque River based on the DENR guidelines, accounting for the time and location of sampling.

Water on Earth's surface is a finite resource that must be wisely managed because, as was already established, just a small portion of it is consumable [8]. Additionally, water is utilized for drinking, fisheries, farming, industry, and recreation. Each designated use has specific established chemical, physical, and biological standards required to serve the corresponding purpose. To examine these standards, testing the water quality, which can be divided into three categories—water quality for human consumption, industrial and residential use, and environmental water quality—is necessary [9]. Heavy metals, a component of inorganic chemicals that make up a more significant proportion of contaminants in drinking water than organic chemicals, are one parameter heavily examined in water quality analysis [10]. The primary goal of heavy metal removal is to keep heavy metals out of the human body and maintain our aquatic ecosystems. However, most underdeveloped nations are currently confronted with establishing a treatment process for water systems, owing to their limited economic capacity to employ modern heavy metal removal technology [3].

Studies in the Philippines typically assess water quality and contaminants in a short span. Monitoring is necessary to identify the overall quality and presence of pollutants such as heavy metals. Sampling and access allow for practical evaluation of quality and concentrations. Monitoring is done every two months to identify the water body's status, threats to water use, and the effects of contaminated water. This study focuses on the Parañaque River, which has not been intensely observed in recent years. Nevertheless, the results of this research will benefit future studies.

This study aims to identify the presence and level of heavy metals in the selected water body – the Parañaque River. It also focuses on determining the water quality index and parameters such as color, pH level, salinity, DO, BOD, TSS, and TDS. The heavy metals and water quality parameters will be evaluated based on environmental guidelines. Additionally, the contents of the water will be tested if it fluctuates depending on the time and location of sampling. The water samples will be tested using an analytical technique, specifically Atomic Absorption Spectrometry (AAS). Statistical analysis will be conducted to determine the concentration of heavy metals in the water and its composition. Lastly, from the given data, this study will provide information about the current condition of the water and the related threats brought by contamination in the river.

The research will benefit the industrial sector of developed cities, such as Parañaque, since it is widely recognized as a first-class, highly urbanized city. Industries must understand and acknowledge the importance of proper wastewater treatment as this will enhance their disposal system that directly goes to the bodies of water. Relative to improving water quality, this research will provide data on how the city can establish a centrally controlled sewage system and develop a treatment for industrial effluent from factories. Making them aware of this problem's seriousness will require them to create a reasonable response to improve water quality continually.

The research monitored the Parañaque River for heavy metals and water quality parameters. Sampling sites were identified, and the analysis was conducted for two months, collecting water samples every two weeks, in the morning and afternoon, when no rain occurred. Water samples were obtained without disrupting the water flow, and Atomic Absorption Spectrometry (AAS) was used to determine the heavy metals present. The data were interpreted by descriptive statistics, ANOVA, Kruskal Wallis, Paired T-Test, and Wilcoxon Signed Rank Test.

2 Methodology

2.1 Water sample sourcing

For this study, water samples were taken from 3 to 6 inches below the surface in various sites of the Parañaque River exposed to wastewater from industries, mainly La Huerta. The shallower depths reduce environmental impact and ensure accuracy, as deeper samples can result in inaccurate representations of the water.

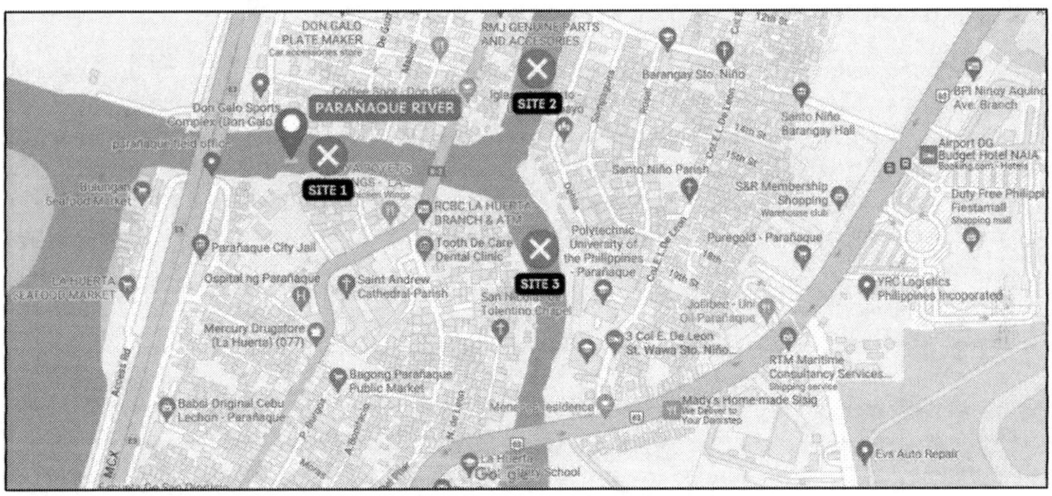

Fig. 1. Sampling sites in Parañaque River.

As shown in Fig. 1, three sampling sites were deliberately chosen as these portions of the river, based on studies, have been identified as most likely to provide valuable data. The sampling sites must be far away from buildings, factories, or other structures to avoid any potential interference and unnatural flow patterns from artificial structures. Furthermore, to ensure the accuracy of the data collected, the samples must be taken at contrasting times of the day, as the behavior of the water can vary greatly depending on the temperature and the presence of industrial activities. The monitoring results were used to determine the condition of the Parañaque River.

2.2 Water sample preparation and collection

Grab sampling is a method of surface water sampling listed in the Water Quality Monitoring Manual Volume 1 of DENR-EMB. It requires sufficient sampling and appropriate site selection to guarantee the reliability and accuracy of assessing water quality parameters, including the presence of heavy metals.

The laboratory facilities required three 1-liter PET bottles filled to the brim for every sampling site. It is essential to wear arm-length powderless nitrile gloves when collecting water samples to avoid touching the container's interior, lip, and caps as much as possible. Before collection, the containers and caps must be rinsed three times and turned downstream to toss the water [11,12]. Be cautious when filling the bottles with samples to avoid any solid matter. Seal the bottles with electrical tape and make sure there are no bubbles. Place the samples in a cooler with ice to keep them at a low temperature and leave them there until they can be transported to the laboratory. However, freezing should be avoided as it can cause significant changes to the quality of the water sample collected. It is also essential to deliver the water samples within the specified period if any parameters have a short holding time [11].

The water samples were sent to BEST Environmental Services and Testing, Corp. in Dasmariñas City, Cavite, which was done within six hours to prevent any changes in the water quality. The other water samples were sent to Mach Union Laboratories, Inc. for exclusive testing of heavy metal concentrations and salinity levels. The researchers will then examine them using data and statistical analysis, employing the Water Quality Guidelines of the DENR as the framework.

2.3 Statistical treatment of data

Data gathered using the AAS and Water Quality Analysis techniques were analyzed through statistical analysis. Water samples from the river were tested multiple times to minimize errors in the results. The statistical analyses utilized in this study were descriptive statistics, ANOVA and Kruskal Wallis, and Paired T-Test and Wilcoxon Signed Rank Test [13]. The software used to analyze the data were Jeffrey's Amazing Statistics Program (JASP), Statistical Package for the Social Sciences (SPSS), also known as IBM SPSS Statistics, and RStudio.

3 Results and discussion

The river's water quality was determined through the use of DENR Administrative Order (DAO) No. 2016-08, also known as Water Quality Guidelines and General Effluent Standards of 2016. This administrative order divides fresh inland waters into five classes (AA, A, B, C, and D) and marine and coastal waters into four categories (SA, SB, SC, and SD) based on the beneficial use of the water. Among these classifications, Parañaque River is under Class C. The analyses used a significance level of 0.05 or 5%. If the p-values exceed the significance level, it will imply no significant difference or relationship between the variables. If it falls below, it means there is a significant difference or relationship between the variables.

https://doi.org/10.1051/e3sconf/202451401004

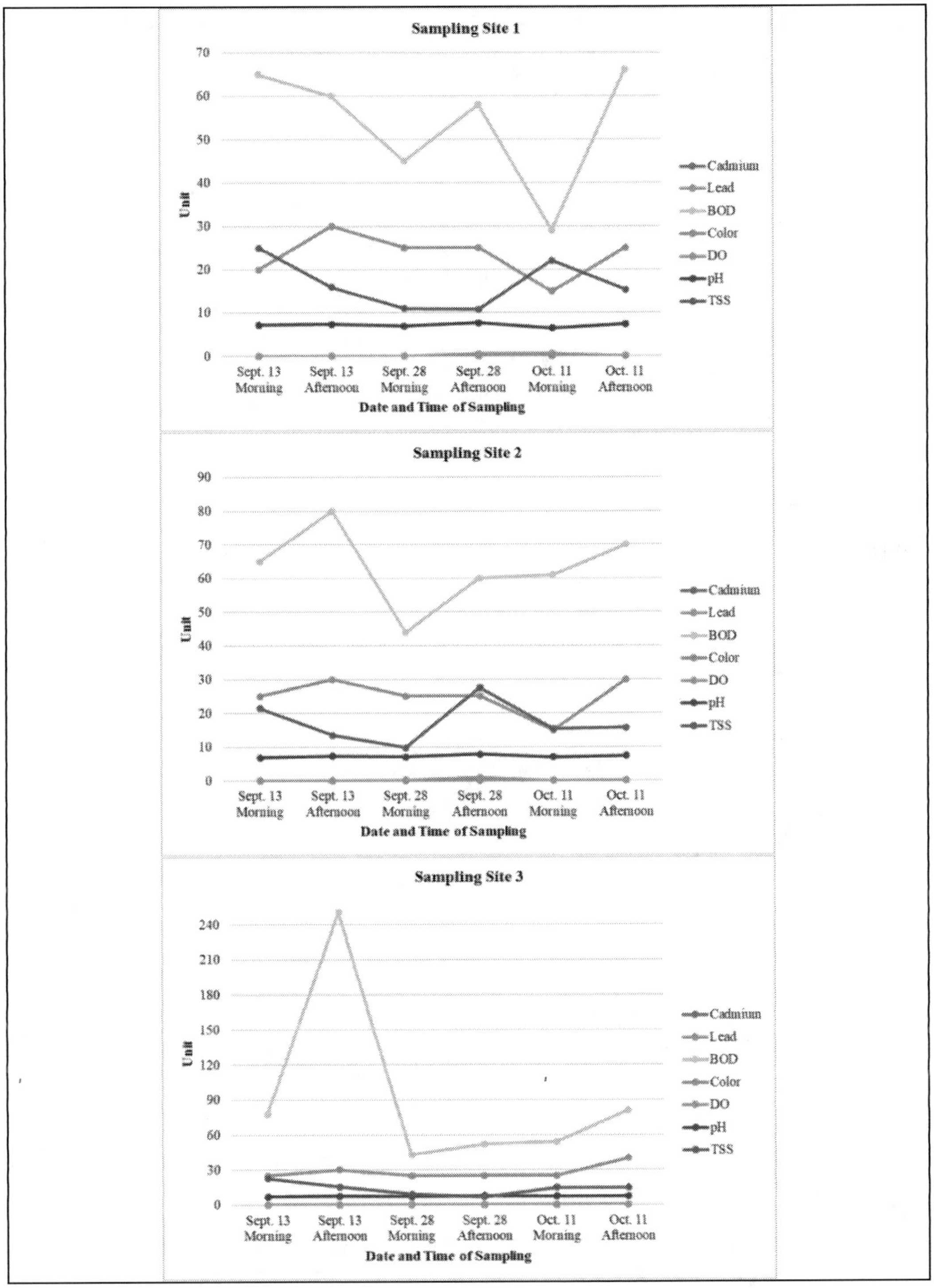

Fig. 2. Comparison of the heavy metal concentrations (Cd and Pb) and water quality parameters across the three sampling sites, which will be examined using DENR guidelines.

Fig. 2 illustrates the comparison of the results obtained from the laboratory facilities. It shows the summary results of the two-month monitoring with two weeks intervals of the water quality in the Parañaque River, particularly the values of cadmium and lead concentrations, BOD, color, DO, pH, and TSS in the three sampling sites considering the time of sampling, morning and afternoon. Salinity and TDS were excluded for separate discussions since these parameters are not included in the guidelines of DENR.

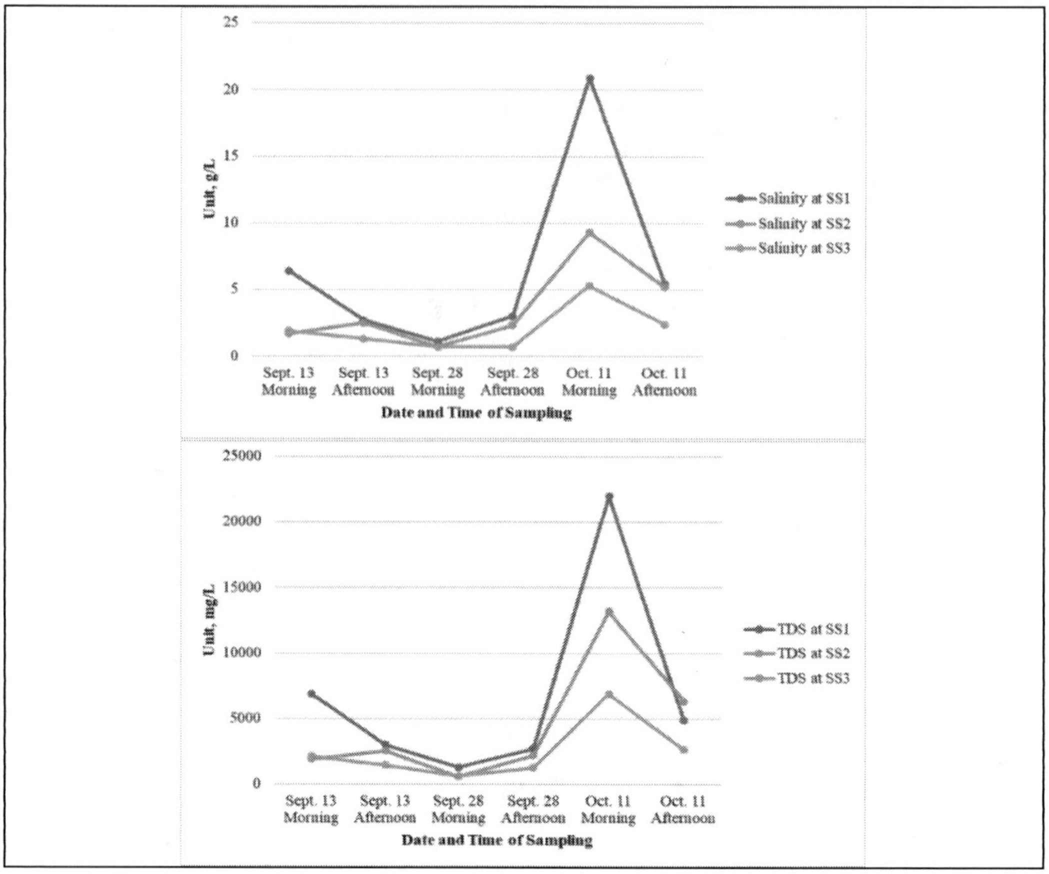

Fig. 3. Comparison of salinity levels and total dissolved solids across the three sampling sites.

As shown in Fig. 3, it shows a graph illustrating the salinity levels at the three sampling sites. The differences in salinity levels between the sampling sites can be attributed to various factors, including the amount of rainwater in the area, the proximity of the sampling sites to the ocean, and the amount of industrial runoff and sewage entering the river. Salinity levels at three sampling sites indicate they are in different river areas. Sampling site 1 is closest to the estuary and has a higher salt concentration with salinity levels ranging from 1.1 to 20.8 g/L. Sampling site 2 is further upstream and has a lower salt concentration with salinity levels ranging from 0.7 to 9.3 g/L. Sampling site 3 is the furthest upstream and has the lowest salt concentration, with salinity levels ranging from 0.7 to 5.3 g/L, likely due to the increasing dilution of salt water with fresh water.

The TDS levels of an urban river vary greatly depending on the location of the sampling sites. The usual range of TDS for rivers is 100 to 20,000 mg/L. Sampling site 1 had the highest levels of TDS, ranging from 1,305 to 22,000 mg/L, sampling site 2 ranged from 595 to 13,190 mg/L, and sampling site 3 ranged from 625 to 6,895 mg/L. This variation is likely due to the proximity of the sampling sites to sources of pollution, the presence of natural sediment, and the concentration of organic matter in the river. Sampling site 1 may be located in an area with higher levels of pollution or sediment accumulation, sampling site 2 may have lower TDS levels due to its location further away from sources of pollution or a higher concentration of natural sediment, and sampling site 3 may have the lowest TDS levels due to its location being further away from sources of pollution and lower concentrations of natural sediment.

Average water sampling results show consistent cadmium and lead concentrations, regardless of the time and location, below 0.002 mg/L and 0.006 mg/L, respectively. Low cadmium and lead concentrations indicate that the water is not contaminated with heavy metals, a positive sign of water quality. It is incredibly encouraging, given that these levels of heavy metals are far below the DENR's safety standards. Moreover, these values suggest that the river is relatively free of toxic substances and that the water quality will likely remain consistent over time.

Table 1. Overall Parañaque River's water condition based on location and time of sampling.

Parameter	Unit	Average Result based on Location of Sampling			Average Result based on Time of Sampling		Decision
		Site 1	Site 2	Site 3	Morning	Afternoon	
BOD, 5-day	mg/L	53.83	63.33	93.17	53.78	86.44	High
Color	TCU	23.33	25.00	28.33	22.22	28.89	Low
DO	mg/L	0.18	0.18	0.13	0.13	0.20	Low
pH	-	7.13	7.25	7.33	6.98	7.49	Normal
TSS	mg/L	16.68	17.22	13.87	16.76	15.09	Low

Table 1 presents the three-week overall comparison of Parañaque River's water condition for three sampled locations in both morning and afternoon. It reveals that the sampling outcomes were consistent across all sites and parameters. The analysis relying on DENR criteria illustrates BOD and DO as low or poor water quality indicators. While color, pH, and TSS revealed an average water quality. A high 5-day BOD indicates a high level of organic matter in the water, which can be caused by human activity or natural sources. It will decrease the oxygen available to aquatic life and reduce the amount of DO in the water. Low DO indicates that the water has insufficient oxygen to support aquatic life, which is generally considered a sign of poor water quality [14].

The Parañaque River has high levels of BOD due to the wastewater treatment plant in La Huerta. This plant discharges organic materials decomposed by microorganisms, contributing to the high BOD levels in the river. Other sources of BOD include leaves and woody debris, dead plants and animals, and animal manure. It will then use oxygen in its process [15]. Although other water quality parameters can be tested to thoroughly examine the cause of high BOD and low DO levels, it is crucial to determine the current condition of the river. Upon surveying the area, much domestic waste is floating in the river. Almost all kinds of trash and waste can be seen in the river, which mainly contributes to the pollution and contamination of the river. The most compromised water quality is notably concentrated in residential areas, primarily due to the untreated or inadequately treated wastewater from local point sources. This untreated wastewater is directly discharged into the surrounding environment, exacerbating the decline in surface water quality within the observed area [16,17].

In India, water pollution primarily stems from the discharge of domestic sewage from towns and cities, with industries also contributing to the issue [18]. The poor water quality is further compounded by inadequate treatment and sewage removal practices. Additionally, deficient urban planning has resulted in the direct disposal of fecal matter into the Ganges River, rendering it one of the most polluted rivers globally. The discharge of untreated urban and domestic waste elevates the concentration of organic matter in the receiving waters [17].

Overall, a combination of different parameters was used to assess water quality. Low levels of cadmium, lead, color, and TSS are generally considered positive water quality indicators. The possible cause of the sudden increase in cadmium and lead concentrations may be due to the industries near the river. As enumerated in the description of Parañaque River, industries such as plastic, pipe, fabric, electronic manufacturing, food packaging, hospitals, and constructions are present around the area. The Parañaque River is connected to Manila Bay, one of the fishing spots in Metro Manila. Specifically, BOD, DO, and TDS levels suggest that the river is biologically dead. Regular water quality monitoring is essential to ensure the health of aquatic ecosystems, provide clean drinking water, and identify potential contamination sources. It is also crucial in determining the safety of fishing areas and preventing diseases caused by consuming contaminated seafood.

Table 2. Comparison of three sampled sites at Parañaque River using Kruskal Wallis and ANOVA.

Parameter	Test	F	H	p	Decision
Cadmium	Kruskal Wallis		2	0.368	Retain H_0
Lead	Kruskal Wallis		2	0.368	Retain H_0
Salinity	Kruskal Wallis		4.028	0.133	Retain H_0
BOD, 5-day	Kruskal Wallis		1.266	0.531	Retain H_0
Color	Kruskal Wallis		1.740	0.419	Retain H_0
DO	Kruskal Wallis		0.066	0.967	Retain H_0
pH	ANOVA	0.531		0.599	Retain H_0
TSS	ANOVA	0.563		0.581	Retain H_0
TDS	Kruskal Wallis		2.667	0.264	Retain H_0

H_0: There is no significant difference in the water quality of the three sampled locations.
H_1: At least one of the sampled locations differs significantly in water quality.

Table 2 exhibits the comparison of three sampled locations at Parañaque River using Kruskal Wallis and ANOVA. Since the p-values of cadmium, lead, salinity, BOD, color, DO, pH, TSS, and TDS are greater than the significance level, retain the null hypothesis. Therefore, there is no significant difference in the water quality of the three sampled locations across all parameters at Parañaque River. It is likely because the water in the river travels upstream and downstream, so contaminants and other factors will be evenly distributed throughout the river.

Table 3. Comparison of morning and afternoon sampling at Parañaque River using Wilcoxon Signed Rank Test and Paired T-Test.

Parameter	Test	t	z	df	p	Decision
Cadmium	The variance is equal to zero					
Lead	The variance is equal to zero					
Salinity	Wilcoxon Signed Rank Test		1.260	8	0.234	Retain H_0
BOD, 5-day	Wilcoxon Signed Rank Test		-2.547	8	0.013	Reject H_0
Color	Paired T-Test	-3.266		8	0.011	Reject H_0
DO	Wilcoxon Signed Rank Test		-0.535	8	0.789	Retain H_0
pH	Paired T-Test	-5.745		8	<.001	Reject H_0
TSS	Wilcoxon Signed Rank Test		1.120	8	0.294	Retain H_0
TDS	Wilcoxon Signed Rank Test		1.244	8	0.250	Retain H_0

H_0: There is no significant difference between Morning and Afternoon sampling.
H_1: There is a significant difference between Morning and Afternoon sampling.

Table 3 shows a significant difference between morning and afternoon values for BOD, color, and pH, with the afternoon values being higher than the morning values. The results are related to the fluctuation in water temperature throughout the day, which varies from morning to afternoon. The sampling is done in the afternoon, which will increase the temperature of the water in the river, causing the metals to dissolve [19]. For BOD, as the temperature increases, its value increases as well. The rate of photosynthesis by the algae and other plant life increases, making it faster to die. Bacteria will then decompose and need oxygen, resulting in a higher demand for oxygen in the area [20]. Rising temperatures can cause an increase in the number of microscopic organisms. Typically, higher temperatures lead to decreased pH values [21], but this was not the case in this sampling, likely due to domestic waste that lowers water quality.

Just like the Mekong River Delta in Vietnam, the Parañaque River faces a critical challenge as it strives to balance the exploitation, rational use, and protection of surface water resources amid the accelerating industrialization and modernization of the country, resulting in increased concentrations of pollutants that render the surface water unsuitable for various purposes [16,17].

4 Conclusion

This experimental study aimed to monitor the water quality in the Parañaque River. The results of the analyses were tabulated and compared to the acceptable levels from the DAO No. 2016-08. Cadmium and lead were detected in the water with less than 0.002 mg/L and less than 0.006 mg/L, respectively. Even though the values can be considered negligible, it is essential to address the heavy metal concentrations before it gets high in the succeeding years. The BOD, DO, and TDS exceeded the acceptable limits of DENR. Meanwhile, color and TSS were lower than the permissible limits, suggesting it is better than the indicated standard. Salinity and pH are all within the acceptable range as well. Using Kruskal Wallis and ANOVA, there is no significant difference in the water quality of the three sampling sites. Using Wilcoxon Signed Rank Test and Paired T-Test, BOD, color, and pH showed a significant difference between morning and afternoon sampling. Conversely, DO, TSS, and TDS displayed no significant difference between morning and afternoon sampling.

Therefore, the research findings reveal that the water quality of the Parañaque River did not generally meet the requirements of the DENR for Class C or Recreational Water Class II waters from September to October due to its poor water quality. The river tested positive for high levels of BOD and low DO levels. Consequently, Parañaque River did not pass the standards set by the DENR and is unsuitable for recreational water usage. The river's poor water quality is because it is heavily polluted by domestic and industrial waste. In addition, the Parañaque River is also affected by bacteria and other microorganisms. The presence of these microorganisms makes the water unsuitable for recreational activities. Additionally, the river's water quality can deteriorate over time if the pollution levels are not addressed and effectively managed. Thus, it is recommended that the local government take immediate action to reduce the pollution levels to make the river fit for recreational water use.

This research will serve as a guide for future studies regarding water quality and heavy metal contamination in the Parañaque River. Furthermore, this study can be used in the future as a basis for monitoring the water quality parameters in the said river or other related bodies of water. It will then be easier to determine the necessary actions to improve the water quality, such as water treatment. Another possible course of this study is to ensure proper wastewater disposal in the river by keeping an eye on the different industries present along the river, as it may affect the livelihood of the people within the area who utilize the river (mainly fishing). Lastly, additional water quality parameters, broader sampling points, and a longer monitoring period can be observed to assess the condition of the Parañaque River as a whole.

References

1. Prof. A. Balasubramanian 2015 (PDF) The World's Water *Cent. Adv. Stud. Earth Sci. Univ. Mysore, Mysore*
2. Thimmaraju M, Sreepada D, Babu G S, Dasari B K, Velpula S K and Vallepu N 2018 Desalination of Water *Desalination and Water Treatment* (IntechOpen)
3. Alidadi H, Tavakoly Sany S B, Zarif Garaati Oftadeh B, Mohamad T, Shamszade H and Fakhari M 2019 Health risk assessments of arsenic and toxic heavy metal exposure in drinking water in northeast Iran *Environ. Health Prev. Med.* **24** 1–17
4. Balali-Mood M, Naseri K, Tahergorabi Z, Khazdair M R and Sadeghi M 2021 Toxic Mechanisms of Five Heavy Metals: Mercury, Lead, Chromium, Cadmium, and Arsenic *Front. Pharmacol.* **12** 227
5. Rahman Z and Singh V P 2018 Assessment of heavy metal contamination and Hg-resistant bacteria in surface water from different regions of Delhi, India *Saudi J. Biol. Sci.* **25** 1687–95
6. Suruchi and Khanna P 2011 Assessment of heavy metal contamination in different vegetables grown in and around urban areas *Res. J. Environ. Toxicol.* **5** 162–79
7. Sankpal S T and Naikwade P V 2012 Heavy metal concentration in effluent discharge of pharmaceutical industries *Sci. Res. Report.* **2** 88–90
8. Roy R 2018 An Introduction to water quality analysis *ESSENCE – Int. J. Environ. Rehabil. Conserv.* 94–100
9. EPA 2021 What are Water Quality Standards? *Stand. Water Body Heal.*
10. Rahmanian N, Hajar S, Ali B, Homayoonfard M, Ali N J, Rehan M, Sadef Y and Nizami A S 2019 Analysis of Physiochemical Parameters to Evaluate the Drinking Water Quality in the State of Perak, Malaysia *J. Appl. Sci.* **2015** 11
11. Danielson T J 2014 Protocols for Collecting Water Grab Samples in Rivers , Streams , and Freshwater Wetlands *Int. J. Environ. Res. Public Health* **26** 6442–71
12. Environment A 2006 *Aquatic Ecosystems Field Sampling Protocols*
13. Qishlaqi A and Moore F 2007 Statistical Analysis of Accumulation and Sources of Heavy Metals Occurrence in Agricultural Soils of Khoshk River Banks, Shiraz, Iran *J. Agric. Environ. Sci* **2** 565–73
14. Higgins P 2013 Biochemical Oxygen Demand - BOD testing and measurement in water *YSI*
15. US EPA, OW 2012 5.2 Dissolved Oxygen and Biochemical Oxygen Demand *Water Monit. Assess.*
16. Giao N T 2022 Determining Key Factors Affecting Surface Water Quality in Soc Trang Province, Vietnam *Int. J. Environ. Sci. Dev.* **13** 144–9
17. Giao N T and Mi L T D 2023 Variation of Surface Water Quality in Cu Lao Dung Island, Soc Trang Province, Vietnam *Int. J. Environ. Sci. Dev.* **14** 202–11
18. Babu M M, Reddy B D, Shivapur A V., Ranjith S, Reddy P N, Kavyateja B V. and Kumar K 2023 A Water Quality Assessment and Modelling of the Indian River Tungabhadra *Int. J. Environ. Sci. Dev.* **14** 160–9
19. Aledia M R G G, Bumanglag A, Carballo G V. and Del Rosario F L U 2019 Removal of Lead (Pb) and Chromium (Cr) in Pasig River by Biosorption using Pineapple (Ananas comosus) Crown

20. Ullah Bhat S and Qayoom U 2022 Implications of Sewage Discharge on Freshwater Ecosystems *Sewage - Recent Advances, New Perspectives and Applications* (IntechOpen)
21. Fondriest Environmental I 2014 Water Temperature - Environmental Measurement Systems *Http://Www.Fondriest.Com* 51–3

Float-Type Photocatalyst Made of Carbon Nitride Microtubes for Environmental Antibacterial Applications

Kunya Nanakida[1], Trang Nakamoto[1] and Kozo Taguchi[1]*
[1]Department of Electrical and Electronic Engineering, Ritsumeikan University, 1-1-1 Noji-higashi, Kusatsu, Shiga, Japan

Abstract. Float-type carbon nitride microtubes (C3N4MT) were synthesized by a calcined hydrothermal process of melamine. Float-type C3N4MT was evaluated by SEM, TEM, and XRD. The XRD showed that float-type C3N4MT was similar to graphitic carbon nitride (g-C3N4) created by calcining melamine. The SEM images revealed that float-type C3N4MT had a variety of tube structures of various sizes. In addition, the TEM images showed that the surface of float-type C3N4MT was covered with a porous structure. Because of its tubular and porous structure, its density was lighter; consequently, it could float on water. Float-type C3N4MT was placed in a liquid microbial growth medium (LB medium) mixed with soil bacteria inoculum to investigate its photocatalytic antibacterial performance on soil bacteria. Float-type C3N4MT was found to reduce the growth of soil bacteria by 15%. Also, float-type C3N4MT can receive more sunlight energy by floating on the water's surface, giving it an advantage in antibacterial activity in muddy water. Therefore, the synthesized float-type C3N4MT can be used for environmental antibacterial applications.

1 Introduction

Nowadays, at least 1.7 billion people worldwide do not have access to clean water [1]. In addition, approximately 300,000 children die each year from diarrhea caused by contaminated water [2]. One source of water pollution is bacteria found in the soil, such as Clostridium perfringens, Bacillus cereus, Staphylococcus, and Escherichia coli. These bacteria can cause food poisoning if consumed and, in some cases, death [3]. Water is commonly disinfected by chlorine or ultraviolet (UV) irradiation. However, chlorine disinfection is not environmentally friendly because it produces trihalomethane, a carcinogen, as a disinfection byproduct [4]. UV disinfection cannot disinfect large amounts of water, and some bacteria are hardly killed by UV light [5]. Therefore, a lot of research has been done to produce clean and safe water today.

Photocatalysts have a strong antimicrobial function because they convert light energy into chemical energy, which exerts an extreme oxidizing power to decompose various

*Corresponding author: n-trang@fc.ritsumei.ac.jp

organic substances. Among them, titanium oxide (TiO2) has attracted wide attention in the field of photocatalysis due to its strong oxidation ability [6]. However, pure TiO2 can only react with light in the UV band, making maximum use of sunlight impossible. To solve this problem, precious metals are loaded on TiO2 so that they can react in the visible light band (Ag, Cu, etc.) [7,8]. However, Ti and other precious metals can lead to water pollution if discharged into water [9]. Ag and Cu are not suitable for antimicrobial use in water because they become toxic ions [10]. Therefore, photocatalytic antimicrobials with safer substances are required.

Graphitic carbon nitride (g-C3N4) is a metal-free photocatalyst composed of ubiquitous elements. g-C3N4 has excellent chemical stability, is non-toxic, and can be prepared at low cost. g-C3N4 has excellent chemical stability, is non-toxic, and can be prepared cheaply, making it suitable for photocatalytic antibacterial applications in water [11]. However, its low conductivity, high recombination frequency of photogenerated carriers, and small surface area limit its use. One way to improve the challenges of g-C3N4 is to tube it [12]. The tubular structure has a large surface area and provides an efficient pathway for charge transfer. Various methods have been used for tubing g-C3N4 today [12–14]. Among them, hydrothermal treatment is simple and inexpensive.

In this study, we synthesized water-floating g-C3N4 microtubes (float-type C3N4MT) by hydrothermal treatment of g-C3N4 and investigated their antimicrobial activity against soil bacteria using light energy. Antimicrobial tests showed a 0.65-fold reduction in bacterial growth. Photocatalytic antimicrobial activity is caused by the decomposition of organic matter in the LB solution, which serves as food for the bacteria, and by the destruction of cell membranes by strong oxidative forces. It is aimed at applications to prevent infectious diseases, as it can be sprayed during flooding to control bacteria in the water.

2 Materials and methods

2.1 Synthesis of g-C3N4

Synthesis of g-C3N4 was performed using the following method. 6g of melamine (Wako Pure Chemical Industries, Ltd, Osaka) was placed in an alumina crucible with a lid to prevent sublimation. The sample was heated to 500°C at a heating rate of 20°C/min for 2 h. Continuously, the sample was heated to 520°C at a heating rate of 5°C/min for 2 h. Finally, the sample was cooled to room temperature and powdered in an agate mortar.

2.2 Synthesis of Float-type C3N4MT

The preparation of Float-type C3N4MT is shown in Fig. 1. Float-type C3N4MT was obtained by hydrothermal treatment of g-C3N4 as follows. 2 g of g-C3N4 was suspended in 60 mL pure water and stirred with a magnetic stirrer for 24 h. The suspension was then autoclaved for 16 h by heating it to 200°C (5°C/min). After the reaction, the autoclave was cooled to room temperature. The supernatant of the product was discarded, and the residue was washed several times with deionized water to remove ammonia. The resulting product was dried at 60°C. The product was then placed in an alumina crucible with a lid and heated in an electric furnace at 5°C/min to 500°C for 2 h. The temperature was further increased to 520°C at 5°C/min and fired for 2 h. The final product was obtained by naturally cooling it to room temperature.

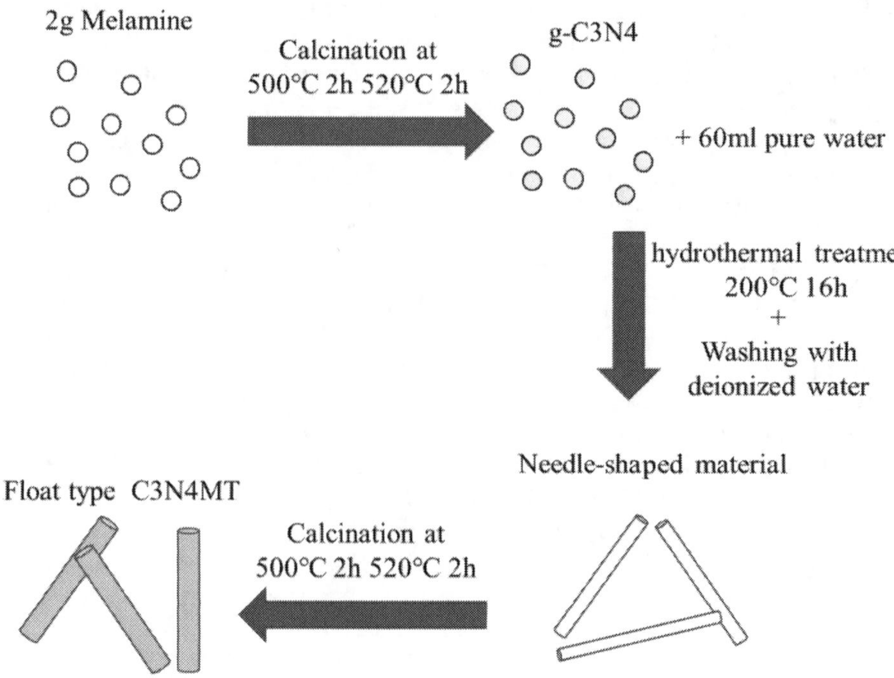

Fig. 1. How to create Float-type C3N4MT.

2.3 Characterization

The phase structure and crystallinity of Float-type C3N4MT were investigated by X-ray diffraction (XRD, X'Pert PRONPD, PANalytical, Ltd, UK) and compared with g-C3N4. The morphology of Float-type C3N4 was observed by scanning electron microscopy (SEM, S-4300, HITACHI, Japan), and the surface structure of Float-type C3N4MT was observed by transmission electron microscopy (TEM, JEM-2100Plus, JEOL, Japan).

2.4 Antibacterial activity test

The LB medium was prepared as follows: 5 g tryptone (Thermo Fisher Scientific Inc, Massachusetts), 2.5 g yeast extract (Thermo Fisher Scientific Inc, Massachusetts), and 5 g sodium chloride (Wako Pure Chemical Industries, Ltd, Osaka) were added to 400 ml deionized water and mixed. After that, 6 ml of 0.32 M NaOH (Wako Pure Chemical Industries, Ltd, Osaka) solution was added, and deionized water was added to obtain the total solution of 500 ml. The solution was autoclaved and stored in the refrigerator for one day. Fig. 2 is an experimental diagram. In a bottle with a volume of 20 ml, 10 ml of deionized water, 1 ml of LB medium, and 1 ml of mud water (collected from a paddy field 34°59'42.7986" N, 135°57'16.1892" E (34.9995222, 135.954497), Shiga, Japan) were mixed. Float-type C3N4MT was then put in and left under sunlight. 1 mL of the solution was taken out daily, and the bacterial cell concentration was measured by evaluating the absorbance at 600 nm in a spectrophotometer (Colourwave CO7500 Colorimeter, Funakoshi, Japan). After that, the measured solution was returned to the same bottle. A diagram of the bacterial concentration is shown in Fig. 3. The equation for calculating the absorbance is Eq. (1).

$$C=log(I/I_0) \quad (1)$$

(C: absorbance, I_0: Incident light intensity, I: Intensity of transmitted light)

https://doi.org/10.1051/e3sconf/202451401005

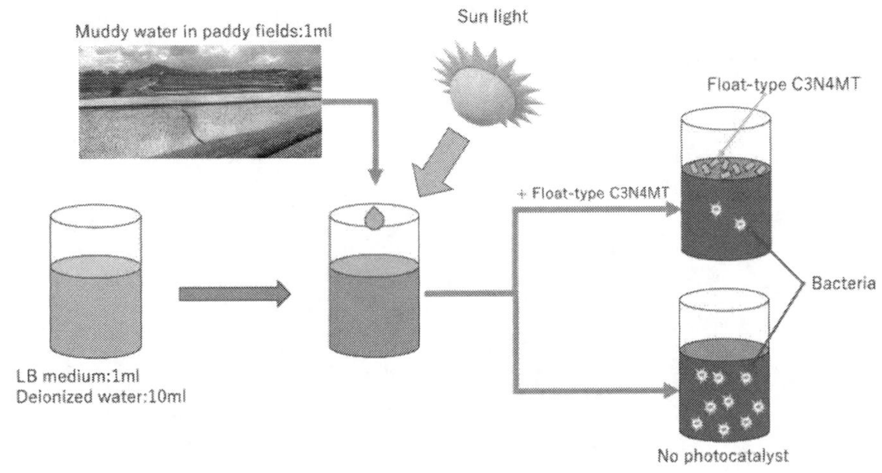

Fig. 2. Antibacterial activity test.

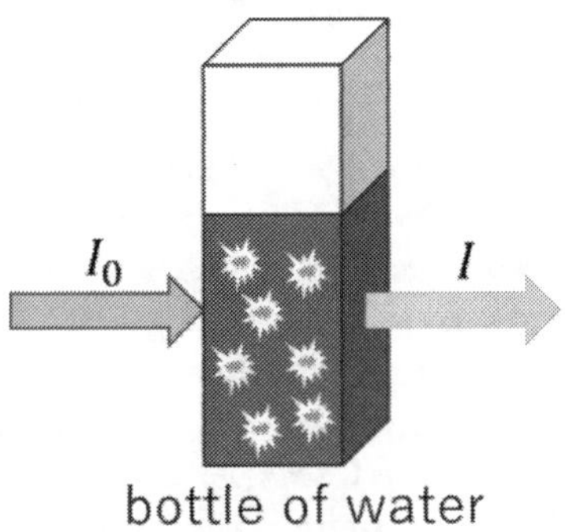

Fig. 3. Absorbance measurement

3 Results and discussion

3.1 Characterization

XRD patterns of g-C3N4 and Float-type C3N4MT are shown in Fig. 4. All samples are found to have two peaks at 13.08° and 27.41°. The small peak at 12.85° corresponds to an interplane distance of d=0.54. The most prominent peak at 27.27° is attributed to being aromatic. g-C3N4 and Float-type C3N4 have peaks in the same phase, indicating that they are the same substance. This result may be due to the wall thickness of Float-type C3N4MT being thinner than that of g-C3N4. The thin walls of the Float-type C3N4MT are another reason it floats on water.

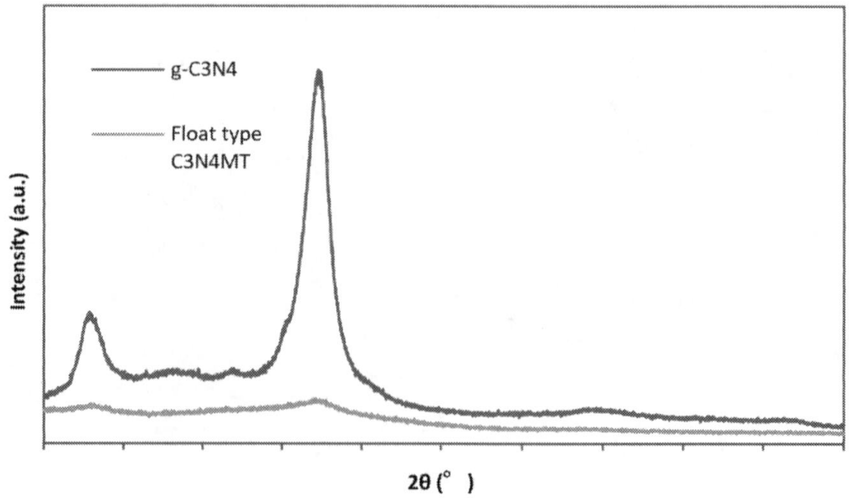

Fig. 4. XRD spectra of g-C3N4 and Float-type C3N4MT.

SEM images of g-C3N4 and Float-type C3N4MT are shown in Fig. 5. Fig. 5(a) is g-C3N4, showing agglomerations of particles of various sizes with irregularities ranging from 1 μm to 10 μm. Fig. 5(b) is Float-type C3N4MT, which has a tubular structure with a diameter of 20 μm and a rod-like structure with a size that can be seen with the naked eye (Fig. 5(c)). From these photographs, it is possible to convert g-C3N4 from particles to microtubes by hydrothermal treatment and calcination. Fig. 5(d) is a TEM image of Float-type C3N4MT. The wall surface of the tubular structure shows that the material is porous, like a sponge. Fig. 5 (c) shows a picture of Float-type C3N4MT after being left in water for ten days. Due to its smaller density, the porous structure allows it to float in water. Therefore, it is possible to perform photocatalytic antibacterial activity without considering the reduction of light energy due to light scattering in water.

Fig. 5. SEM images of g-C3N4 (a) and Float-type C3N4MT (b). A photograph was taken ten days after float-type C3N4MT was placed in water. TEM image of Float-type C3N4MT.

3.2 Antibacterial activity test

The growth curves of soil bacteria in the solution without and with Float-type C3N4MT photocatalyst are shown in Fig 6. The addition of Float-type C3N4MT inhibited the growth of soil bacteria. Soil bacteria were most abundant on day 3 and then decreased as food was lost. The absorbance on day 3 was about 35% smaller in the solution with Float-type C3N4MT than in the solution without a photocatalyst. This result indicates that the photocatalyst converts light energy into chemical energy and has strong oxidizing power, which decomposes bacteria and LB medium in the solution and inhibits their growth.

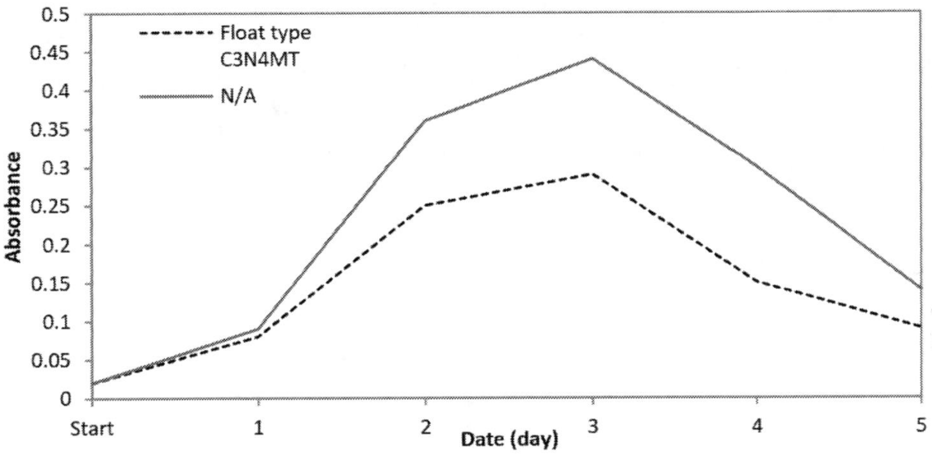

Fig. 6. Growth curves of mud with and without Float-type C3N4MT.

4 Conclusion

In this study, we created a water-floating photocatalyst that exhibits antimicrobial activity in water using light energy and demonstrated its performance. Float-type C3N4MT was created by hydrothermal treatment of g-C3N4, a visible light-responsive photocatalyst composed of ubiquitous elements. XRD confirmed that Float-type C3N4MT is like g-C3N4. Float-type C3N4MT was found to be thinner than g-C3N4. TEM and SEM images showed a tubular and porous structure with a larger reaction area and better conductivity than g-C3N4. Furthermore, its smaller density than water allows it to float, making it more exposed to light to convert light energy into chemical energy efficiently. In antimicrobial tests, the growth of soil bacteria was suppressed, and Float-type C3N4MT may reduce the number of infectious diseases caused by muddy water. Therefore, it can be used to reduce infections during flooding, as it can be easily sprayed and used as an antimicrobial material.

References

1. UNICEF Homepage, https://www.unicef.or.jp/news/2017/0146.html, last accessed 2017/7/12
2. UNICEF Homepage, https://www.unicef.or.jp/news/2015/0348.html, last accessed 2015/11/19
3. A.A. Mostafa, A.A. Al-Askar, K.S. Almaary, T.M. Dawoud, E.N. Sholkamy, M.M. Bakri, Saudi J Biol Sci. **25**. 361-366 (2018)
4. Y. Du, X.T. Lv, Q.Y. Wu, D.Y. Zhang, Y.T. Zhou, L. Peng, H.Y. Hu, J Environ Sci (China). **58**. 51-63 (2017)

5. L. Sun, T. Du, C. Hu, J. Chen, J. Lu, Z. Lu, H. Han, ACS Sustain Chem Eng. **5**. 8693 8701 (2017)
6. H.M. Yadav, J.S. Kim, S.H. Pawar, Korean Journal of Chemical Engineering. **33**. 1989-1998 (2016).
7. G.Y. Nigussie, G.M. Tesfamariam, B.M. Tegegne, Y.A. Weldemichel, T.W. Gebreab, D.G. Gebrehiwot, G.E. Gebremichel, Inter. J. Photoenergy. 2018. 1-7 (2018).
8. B. Wu, R. Huang, M. Sahu, X. Feng, P. Biswas, Y.J. Tang, Science of the Total Environment. **408** (7). 1755-1758 (2010)
9. X. Zhu, Y. Chang, Y. Chen, Chemosphere. **78** (3). 209-215 (2010)
10. O. Bondarenko, K. Juganson, A. Ivask, K. Kasemets, M. Mortimer, A. Kahru, Arch Toxicol. **87** (7). 1181-1200 (2013)
11. F. He, Z. Wang, Y. Li, S. Peng, B. Liu, Appl Catal B. **269**. 118828 (2020)
12. J. Yang, Y. Liang, K. Li, G. Yang, K. Wang, R. Xu, X. Xie, Appl Catal B. **262** 118252 (2020)
13. M. Wu, J. Zhang, B. bei He, H. wen Wang, R. Wang, Y. sheng Gong, Appl Catal B. **241**. 159-166(2019)
14. Z. Jiang, X. Zhang, H.S. Chen, X. Hu, P. Yang, Chem. Cat. Chem. **11**(18). 4558-4567 (2019)

E3S Web of Conferences **514**, 02001 (2024) https://doi.org/10.1051/e3sconf/202451402001
ICERE 2024

Analysis of the technical feasibility of ecological bricks based on PET plastic and bottle glass for the reduction of solid waste pollution in a metropolis

Jessica Arguedas[1*], and *Carlos Eyzaguirre*[2]

[1]Student, Peruvian University of Applied Sciences, Peru
[2]Professor, Peruvian University of Applied Sciences, Peru

Abstract. Contamination by non-biodegradable solid waste is worrying, mainly due to PET and glass waste, because they are used for single use such as bottles or containers. In addition, global production in 2021 of these materials was 139 million tons and 23.4 million tons, respectively. Therefore, the objective of this research is to develop an efficient design of bricks composed of cement, PET plastic and glass, evaluating both their physical properties and their mechanical properties to reduce said waste. For this, 4 mixtures based on cement, sand and glass were designed with proportions of 1:2:0.5 and 1:2:1. In addition, 5% and 10% PET were incorporated into these proportions. It should be noted that the PET was ground to a size of 0.2 cm and the glass was ground to a size of 2 cm. The tests carried out show that the 1:2:0.5 dosage with 10% PET is the most efficient, since it obtained a compressive strength of 77.99 kg/cm^2 after 28 days. Therefore, it exceeds the minimum value of 70 kg/cm^2 after 28 days to be classified as type II brick according to the masonry standard E.070. Likewise, said unit complied with the warping, dimensional variation and absorption tests required by the aforementioned standard.

1 Introduction

Non-biodegradable solid waste is waste that cannot decompose on its own. Therefore, these materials persist for many years in the environment; Therefore, its accumulation generates environmental pollution and harms the living beings that reside in its surroundings.

In addition, most of the elements are inorganic and synthetic such as polyethylene terephthalate (PET) plastic and bottle glass. It should be noted that these wastes are the most used by society, since they are single-use waste, for example, in bottles and containers.

On the one hand, according to the Minderoo Foundation [1] it was noted that in 2021, 139 million metric tons of single-use plastic were produced worldwide. What's more, in Peru, according to the Ministry of the Environment (MINAM) [2] it mentions that 30 kg of plastic is used per inhabitant per year.

* Corresponding author: u201811993@upc.edu.pe

On the other hand, according to FEVE [3] they emphasize that the production of glass for bottles and jars worldwide in 2021 was 23.4 million tons. Likewise, in Peru, according to the Asociación Recíclame Perú [4] they emphasize that, in 2021, 260 thousand tons of glass were used, of which only 25% were made from recycled glass.

Therefore, this article is about the design of an ecological brick based on PET and soda-calcium bottle glass to reduce the environmental impact generated by the non-biodegradable solids, mentioned above.

Numerous investigations have been carried out to evaluate the possibility of adding waste soda-lime glass and PET plastic in the manufacture of concrete bricks and fired clay-based bricks. It is important to consider that such bricks have a high amount of embodied energy and a considerable carbon footprint.

On the one hand, polyethylene terephthalate (PET), its use is ephemeral, so it is discarded. Therefore, large quantities of them end up in landfills, drains and even the ocean. With the above, Dadzie et al. [5] mention that the construction sector has been interested in using this element in materials such as sustainable masonry units, since there are various investigations that used PET that indicate that its obtaining is easier and more accessible, and that it is a good thermal insulator.

There are several studies that applied the PET material with clay or concrete, firstly, Daraman et al. [6] proposed designing ecological bricks based on cement, sand and PET with proportions of 2.5, 5 and 7.5 percent by weight, whose size was less than 5 mm like that of sand. Likewise, the authors evaluated which brick had lower thermal conductivity and greater compressive strength. Therefore, in the results they discovered that the mixture with 5% PET obtained a thermal conductivity of 0.581 W/mK, a compressive strength of 3.90 MPa and a weight of 2.15 kg/m3. It should be noted that its compressive strength complies with what ASTM C129-11 indicates that it must be greater than 3.45 MPa and its weight greater than 2.00 kg/m3 for non-load-bearing bricks.

Second, Akinyele et al. [7] proposed designing clay bricks with PET percentages of 0, 5, 15 and 20% for each unit, whose additive size was between 1 mm to 8 mm. It should be noted that it was observed that units with a percentage greater than 5% of PET disintegrated in the oven due to high temperatures, since its melting point is 250 °C. Therefore, the most efficient unit was 5% PET and 95% clay, because it obtained a compressive strength of 2.30 MPa.

Third, Limami et al. [8] analyzed a comparison between unfired clay bricks based on PET and HDPE for each unit. It is worth mentioning that the additives were ground to sizes less than and equal to 1mm, between 1 to 3 mm and between 3 to 6 mm. Furthermore, the clay was mixed with the additives by melting at 300°C. Of which the most competent units were 3% HDPE and 3% PET with respect to the weight of clay with a size less than 1 mm, since they obtained 4.61 MPa and 4.50 MPa of compressive strength, respectively.

On the other hand, studies have also been investigated in which various authors have used glass with clay and cement for masonry units. First, Malek et al. [9] developed glass-cement composite bricks (CGCB) with an internal printed polyethylene terephthalate glycol (PET-G) scaffold. This innovative material uses 86% waste (78% recycled glass and 8% recycled PET-G), making it an economical and sustainable option. Test results revealed significant improvements in thermal properties, with a 5% increase in thermal conductivity and an 8% decrease in thermal diffusivity and specific heat. Furthermore, the use of the internal scaffold reduced the anisotropy of the mechanical properties of the CGCB. These findings highlight the importance of this approach in the search for efficient and environmentally friendly alternatives in construction.

Secondly, Crespo-López and Cultrone [10] analyzed the addition of 20% by weight of crushed domestic glass to clay soils to manufacture extruded and handmade bricks fired at 800, 950 and 1100° C. The tests carried out revealed that the addition of glass gives the bricks greater compactness, strength and durability than those without glass fired at the same temperature. As a result, it was determined that 20% by weight is the ideal proportion for glass waste added to the clay used in the production of high-quality compact bricks.

Third, Xin et al. [11] made clay bricks with the addition of waste glass from beer bottles and were fired at a temperature between 900°C and 1050°C. It was determined that the maximum compressive strength of 113.93 MPa was achieved with the addition of 10% waste glass at a firing temperature of 1050°C.

Likewise, Hasan et al. [12] made bricks with 2%, 4%, 10%, 16%, 30% and 40% waste glass waste instead of natural clay. The results obtained indicated that bricks prepared from the mixture of clay and 30% soda-lime glass waste are the best option, since it shows a 41% increase in compressive strength compared to clay bricks.

The main purpose of this article is to analyse the results of the performance properties of the design of a new ecological brick based on PET and soda-lime glass with the aim of evaluating which mix design is more efficient.

2 Methodology

2.1 Materials

The sustainable brick is composed of ground PET material (Fig. 1a), crushed soda-lime glass (Fig. 1b) from beer bottles, Type I Portland cement (Fig. 1c) and fine aggregate (Fig. 1d).

The collection of waste PET plastic and soda-lime glass from beer bottles was done with the help of family and friends, as they stored their bottles so they could be recycled. Likewise, the services of the company Reyes Inversions were hired for the grinding of PET plastic, whose particle size was 0.2 cm and an apparent density of 620 kg/m3. Regarding the crushing of waste glass, a size of 2 cm was achieved. In addition, type I Portland cement, coarse sand and water were used to make the cement brick. It should be noted that the water and cement ratio was 0.5.

2.2 Mix design

In total, 40 masonry units were made with dimensions of 13 x 22 x 10 cm. For the mix design, two proportions of cement, sand and crushed glass were made, which were used for the dosage of 1:2:0.5, shown in Table 1, and for the dosage of 1:2:1, it is presented in Table 2.

https://doi.org/10.1051/e3sconf/202451402001

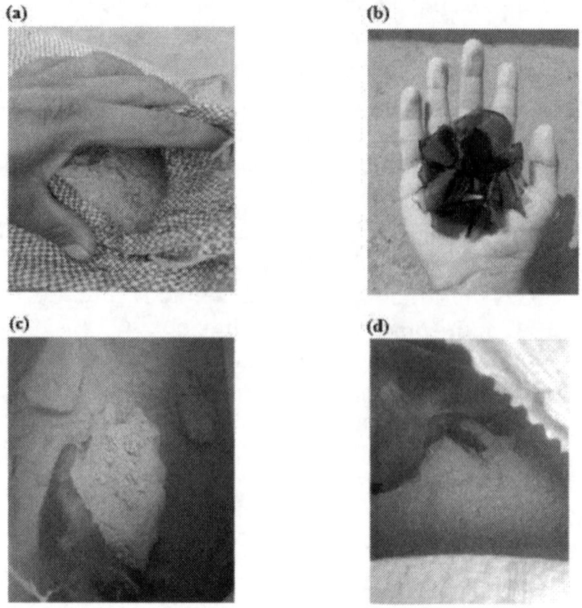

Fig. 1. Materials used for the production of ecological brick. (a) Ground PET, (b) Crushed glass, (c) Portland cement type I, (d) Coarse sand.

In addition, 5% and 10% PET were added to both dosages, so 4 mixture designs were prepared.

2.3 Experimental tests

In relation to the standards and regulations applied to the design of ecological brick, it is evident that the requirements of the tests to be carried out must comply with the minimum established by Masonry Standard E.070. Likewise, the corresponding Peruvian Technical Standards (NTP) must be followed to guarantee proper execution of each test.

These standards include Warping (NTP 339.613), Dimensional Variation (NTP 339.613 and NTP 339.604), Compression Strength (NTP 399.613 and NTP 339.604) and Absorption (NTP 339.613) tests.

2.4 Preparation of specimens

Firstly, PET and glass waste discarded in landfills and/or collection centers were collected. Then, these wastes were washed and dried in the open air for a period of 4 days. The PET was then crushed to a particle size of 0.2 cm, while the waste glass was crushed to a size of 2 cm. In addition, a Lego-type mold with dimensions of 13 x 22 x 10 cm was designed to shape the prototypes.

Table 1. Dosage 1:2:0.5

N° Mixture	W/C Ratio	Cement (kg)	Sand (kg)	Glass (kg)	Water (L)	PET (kg)	PET (%)	Total weight (kg)	N° of units
1	0.5	16.4	32.7	8.2	8.2	1.6	5%	67.1	10
2	0.5	16.4	32.7	8.2	8.2	3.3	10%	68.8	10

Table 2. Dosage 1:2:1

N° Mixture	W/C Ratio	Cement (kg)	Sand (kg)	Glass (kg)	Water (L)	PET (kg)	PET (%)	Total weight (kg)	N° of units
3	0.5	16.4	32.7	16.4	8.2	1.6	5%	75.3	10
4	0.5	16.4	32.7	16.4	8.2	3.3	10%	77	10

Afterwards, the mix design was carried out using two different proportions of cement, sand and crushed glass, these being 1:2:0.5 and 1:2:1 respectively. To these mix designs, 5% and 10% of PET were added with respect to the proportion of sand.

It is worth mentioning that these proportions were determined by reviewing previous research that used PET and glass in concrete and fired brick masonry units. Next, all materials were mixed homogeneously and water was added gradually at a water-cement ratio of 0.5.

Then, the resulting mixture was poured into the previously prepared molds. Subsequently, various tests were carried out as stipulated by the Peruvian Technical Standard E.070 for Masonry, such as compression resistance, absorption, dimensional variation and warping tests, in order to evaluate the efficiency of the different mix designs. Finally, an evaluation will be carried out to determine the number of kilograms of PET plastic and glass used in the creation of the prototypes.

In Fig. 2, the flow chart of the summarized specimen preparation can be seen.

3 Results

3.1 Warping test

Tests were carried out with 10 bricks of each mixture composition. On this occasion, it was evaluated whether each brick had a concave or convex surface on its upper and lower face. According to standard E.070, the difference between concavity and convexity should not exceed 10 mm. The results corresponding to each mixture composition are detailed in Fig. 3.

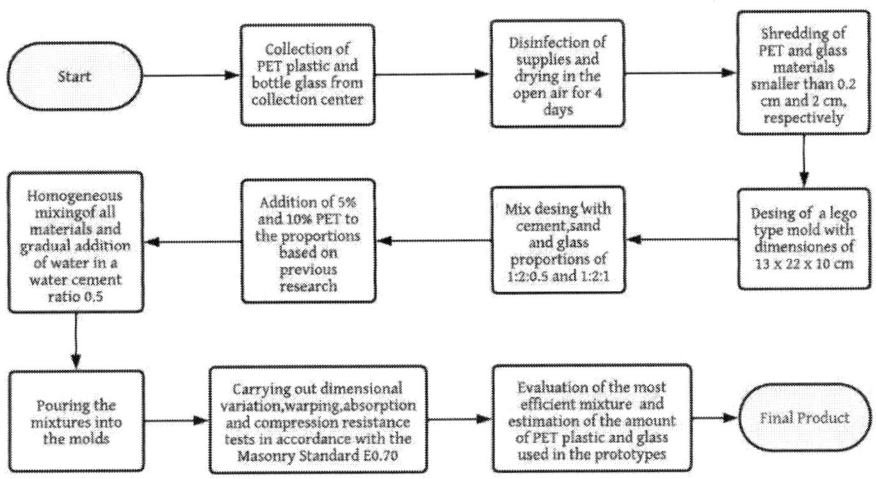

Fig. 2. Flowchart of the production of ecological brick.

 https://doi.org/10.1051/e3sconf/202451402001

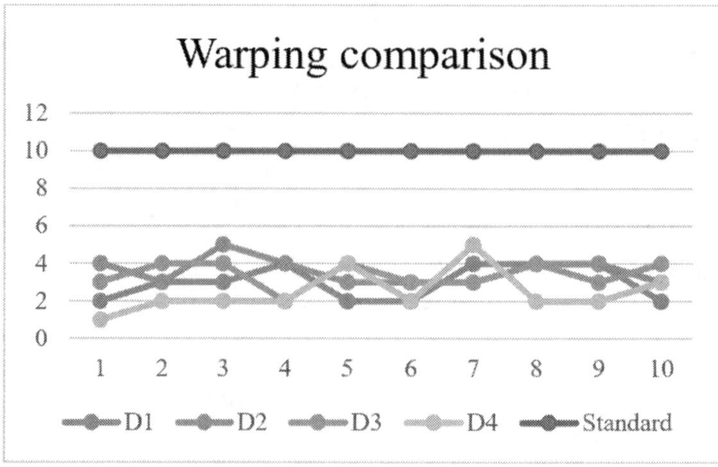

Fig. 3. Comparison of warping test results for different dosages.

Regarding the analysis of the results of the Warping Test, a line graph was used to compare the results of each type of brick. In Fig. 3 it can be seen that the four dosages satisfactorily completed the warping test, given that it is below the maximum allowed, as mentioned previously.

3.2 Dimensional variation

The dimensional variation test was carried out on 10 bricks of composition of the proposed mixtures. In other words, for each dimension the average of these was determined for each composition of the mixture as can be seen in Table 3.

With respect to the previous table, it can be seen that the 4 mix designs comply with what the Masonry Standard stipulates. That is, the results of the dimensional variation of length, height and width do not exceed 4%, 8% and 6%, respectively.

Regarding the length of the brick, dosage 1 obtained the best result (-0.55%), within the limit allowed by the standard of ±4%. Regarding height, dosage 1 obtained the best result (-0.50%), within the allowed limit of ±8%. Finally, in the width dimensions, dosage 3 obtained the best result (-0.54%), within the allowed limit of ±6%.

3.3 Absorption test

In this evaluation, five bricks of each dosage variant were tested, with the aim of obtaining a more complete view of the absorption capacity of the units. According to the Peruvian Technical Standard E.070 for masonry, the absorption percentage should not be greater than 12%. The average absorption was determined for each mixture design and the results are detailed in Table 4.

Table 3. Summary of results of variation of dimensions of the 4 mixtures

Dimensions	Mix Design				Masonry standard
	1	2	3	4	
Long	-2.82%	-2.91%	-0.55%	-1.73%	< ± 4%
High	4.10%	4.40%	-0.50%	4.40%	< ± 8%
Broad	-2.77%	-3.92%	-0.54%	-1.69%	< ± 6%

Table 4. Summary of absorption results of the 4 mixtures.

Mixture design	Absorption (%)	According to NTP E.070: Absorption >12%
D1	10.8	Yes
D2	10.7	Yes
D3	10.3	Yes
D4	9.3	Yes

Table 5. Summary of results of the compressive strength test at 28 days of the 4 mixtures

Mix Design	Average compressive strength (kg/cm^2)
D1	69.30
D2	77.99
D3	46.81
D4	70.34

3.4 Compression strength test

In this test, five bricks of each mixture design were presented to analyze which mixture is more resistant and complies with the E.070 masonry standard. That is, the compressive strength achieved after 28 days of curing must be more than 70 kg/cm2 for the brick to be classified as type II.

In table 5, you can see the compressive strength of the 4 mix designs at 28 days. In this table it can be seen that the dosage of 1:2:0.5 with 10% PET plastic obtains the highest compressive strength with 77.99 kg/cm2.

4 Conclusions

It was determined that the absorption of bricks with a 1:2:1 dosage with an addition of 5% and 10% of PET plastic, were the ones with the lowest percentage of absorption, being 10.3 and 9.3, respectively, which are within the allowed range. Therefore, the use of bottle glass and PET plastic improves certain physical properties of the bricks, such as reducing their water absorption capacity.

It was discovered that the four mixture designs carried out comply with what is specified in the E.070 standard regarding dimensional variation and warping tests. That is, with respect to the first test, the ecological bricks comply with the dimensions of a Type I brick. Likewise, the ecological bricks comply with not exceeding the 10 mm limit of the second test.

The highest average compressive strength at 28 days was mixture 2, which has a 1:2:0.5 ratio of cement, sand and glass with a PET percentage of 10%, obtaining 77.99 kg/cm2, exceeding the value minimum of 70 kg/cm2 to be classified as a type II brick according to the Peruvian standard e0.70 for masonry.

Mix design 2 with respect to the dimensional variation test can be classified as a type II brick, since its dimensions of length, height and width are less than 4%, 7% and 6% as indicated by Masonry Standard E.070. In addition, the design 2 mixture obtained a value of 3.6 in the warping test, thus meeting the maximum value with respect to a type II brick, which is 4 mm.

References

1. Minderoo Foundation, “Plastic Waste Makers Index.” Plastic Waste Makers, 2023.
2. Ministerio del Ambiente (MINAM). “Cifras del mundo y el Perú.” 2020. https://www.minam.gob.pe/menos-plastico-mas-vida/cifras-del-mundo-y-el-peru/ (accessed Jun. 18. 2023).
3. The European Container Glass Federation (FEVE). “EU Container glass industry records highest ever growth in production.” https://feve.org/year2021production/ (accessed Jun. 18. 2023).
4. El Comercio. “En el Perú el 25% de las 260 mil toneladas de vidrio producidas contienen vidrio reciclado.” 2021. https://elcomercio.pe/economia/en-el-peru-el-25-de-las-260-mil-toneladas-de-vidrio-producidas-contienen-vidrio-reciclado-noticia/?ref=ecr (accessed Jun. 18. 2023).
5. D. K. Dadzie. Kaliluthin A.K. and Raj Kumar D.. “Exploration of waste plastic bottles use in construction.” Civil Engineering Journal. 2020. Accessed: Nov. 02. 2022. [Online]. Available: shorturl.at/dgM57I. S. Jacobs and C. P. Bean, “Fine particles, thin films and exchange anisotropy,” in Magnetism, vol. III, G. T. Rado and H. Suhl, Eds. New York: Academic, 1963, pp. 271–350.
6. R. Deraman. M. N. M. Nawi. M. N. Yasin. M. H. Ismail. and R. S. M. O. M. Ahmed. “Polyethylene Terephthalate Waste Utilisation for Production of Low Thermal Conductivity Cement Sand Bricks.” Journal of Advanced Research in Fluid Mechanics and Thermal Sciences. 2021. Accessed: Jun. 20. 2023. doi: 10.37934/arfmts.88.3.117136.
7. J. O. Akinyele. U. T. Igba. and B. G. Adigun. “Effect of waste PET on the structural properties of burnt bricks.” Sci Afr. vol. 7. p. e00301. Mar. 2020. doi: 10.1016/J.SCIAF.2020.E00301.
8. H. Limami. I. Manssouri. K. Cherkaoui. and A. Khaldoun. “Study of the suitability of unfired clay bricks with polymeric HDPE & PET wastes additives as a construction material.” Journal of Building Engineering. vol. 27. Jan. 2020. doi: 10.1016/J.JOBE.2019.100956.
9. M. Małek et al.. “Performance Properties of Cement–Glass Composite Bricks (CGCB) with Additively Manufactured (AM) Polymeric Scaffolding.” Materials 2023. Vol. 16. Page 1909. vol. 16. no. 5. p. 1909. Feb, 2023. doi: 10.3390/MA16051909.
10. L. Crespo-López and G. Cultrone. “Improvement in the petrophysical properties of solid bricks by adding household glass waste.” Journal of Building Engineering. vol. 59. Nov. 2022. doi: 10.1016/J.JOBE.2022.105039.
11. Y. Xin et al.. “Recycling crushed waste beer bottle glass in fired clay bricks.” vol. 11. 2021. Accessed: Nov. 05. 2022. [Online]. Available: shorturl.at/cuw18.
12. M. R. Hasan. A. Siddika. M. P. A. Akanda. and M. R. Islam. “Effects of waste glass addition on the physical and mechanical properties of brick.” Innovative Infrastructure Solutions. vol. 6. no. 1. pp. 1–13. Nov. 2020. doi: 10.1007/S41062-020-00401-Z.

https://doi.org/10.1051/e3sconf/202451402002

Investigating the Optimization of Magnetic Biochar Production from Rubber Seed Shells for Enhanced Cr(VI) Removal Efficiency

Yasumin Prangmoo[1], *Orasa* Choolaaied[1], *Naree* Phanchindawan[1], *Warunee* Limmun[1*], and *Thatchapol* Chungcharoen[1]

[1]King Mongkut's Institute of Technology Ladkrabang, Prince of Chumphon Campus, 17/1, Chumko Pathio, Chumphon, 86160, Thailand

Abstract. This study aimed to utilize agricultural and produce low-cost magnetic biochar from rubber-seed shells using ferric chloride ($FeCl_3$) as a transition metal. The study employs Response Surface Methodology (RSM) based on the Box-Behnken Design (BBD) to determine optimal production conditions for removing chromium (Cr(VI)). The effect of preparation conditions such as pyrolysis temperature (500-700 °C), duration (90-180 min), and impregnation (1-3 M) on the produced magnetic biochar was examined. The optimal condition was demonstrated based on yield percentage and Cr(VI) removal efficiencies. The study revealed that the optimal conditions for producing magnetic biochar from rubber seed shells were a pyrolysis temperature of 580 °C, a pyrolysis time of 130 min, and a $FeCl_3$ concentration of 3 M. Under these conditions, a yield of 48.63% was achieved, and the removal efficiencies for Cr(VI) were 41.29%. This research suggests that utilizing agricultural waste products from rubber seed shells may be a viable and economical method for producing magnetic biochar, which can serve as an efficient adsorption agent.

1 Introduction

Chromium (Cr(VI)) is prevalent in various industrial processes, including electronics, tanning, and metal coating. However, conventional methods for removing it from wastewater are challenging. Cr(VI) is a significant environmental pollutant and poses a considerable public health risk due to its association with cancer and skin disorders [1, 2]. It can also enter the human body through food or inhalation, and its accumulation over time can lead to severe health complications [3, 4]. Consequently, removing Cr(VI) before discharging into the environment is crucial.

Thailand emerged as the largest producer and exporter of natural rubber, having plantations that covered more than 3.9 million hectares in 2022 [5]. Each hectare can yield approximately 150 kg of rubber seeds [6]. However, rubber tree plantations generate substantial waste in the form of rubber seed shells (RSS). Hence, utilizing RSS as a valuable feedstock has the potential to enhance the overall value in Thailand. Chaireh [7] reported that

* Corresponding author: warunee.li@kmitl.ac.th

there has been limited use of RSS in Thailand. However, the disposal of RSS often presents environmental challenges. This research proposes a solution to these challenges by transforming this waste into a resource by producing magnetic biochar.

Magnetic biochar, an absorbent material, is derived from agricultural waste through a specific process. This process involves adding transition metals or oxides, such as ferric chloride ($FeCl_3$), to the waste. Following this addition, the mixture undergoes a slow pyrolysis process under oxygen-limited conditions [8]. The temperature and duration of the pyrolysis significantly influences the properties of the final product, such as surface structure, carbon content, pH, ash content, and stability [9]. These properties make magnetic biochar a potent tool for removing heavy metals from wastewater, as demonstrated in previous studies [10, 11].

Experimental design is important to assess the effects of these parameters on the produced magnetic biochar. Many researchers have used response surface methodology (RSM) to study the interactions between multiple parameters [12, 13]. The RSM is an advantageous tool that can help reduce the number of experimental runs while maintaining a statistically significant amount of data for analysis.

This research aims to optimize the production of magnetic biochar from RSS by investigating the effects of various factors such as pyrolysis temperature, duration, and $FeCl_3$ impregnation. The optimization process utilizes the RSM through the Box-Behnken Design (BBD), which allows the study of interactions between process parameters. The response parameters under consideration are magnetic biochar yield and Cr(VI) removal efficiency. This innovative approach holds the potential to transform RSS, a waste product, into a valuable resource for environmental pollution mitigation.

2 Materials and methods

2.1 Materials

Rubber seed shells (Fig. 1(a)) were collected from the rubber plantations in Chumphon Province, Thailand. These shells were then dried and ground into a powder with a particle size of 150 μm. Potassium Dichromate ($K_2Cr_2O_7$) was also dissolved in purified water to prepare a stock solution with a concentration of 1 g/L. All the chemicals used are of analytical grade (AR Grade).

Fig. 1. (a) Rubber seed shells (b) Magnetic biochar

2.2 Preparation of magnetic biochar

Magnetic biochar was synthesized by activating rubber seed shells with $FeCl_3$ at concentrations ranging from 1-3 M. This mixture was stirred for 8 h at room temperature (25±2 °C) and a speed of 150 rpm. Subsequent pyrolysis was conducted at temperatures between 500-700 °C for durations ranging from 90-180 min, with a heating rate of 5 °C/min. These conditions were set based on the design parameters established by RSM. After pyrolysis, the magnetic biochar (Fig. 1(b)) was cooled and stored in a desiccant bowl. The yield percentage was determined based on Eq. (1).

$$\text{Yield (\%)} = \frac{\text{Weight of magnetic biochar}}{\text{Weight of raw material}} \times 100 \qquad (1)$$

2.3 Design of experiment for magnetic biochar

The study employed the Box-Behnken design of Response Surface Methodology (RSM) to determine the influence of three parameters on the production of magnetic biochar: pyrolysis temperature (X_1), pyrolysis time (X_2), and $FeCl_3$ concentration (X_3). The range of factor levels for these parameters is presented in Table 1. The responses considered for optimization were yield percentage (Y_1), and Cr(VI) removal efficiency (Y_2). The main objective was to predict the optimum conditions for magnetic biochar synthesis from rubber seed shells. The quadratic model representing the relationship between the parameters and responses is given by Eq. (2).

$$Y = \beta_0 + \sum_{i=1}^{3} \beta_i X_i + \sum_{i=1}^{3} \sum_{j=1}^{3} \beta_{ij} X_i X_j + \sum_{i=1}^{3} \beta_{ii} X_i^2 + \varepsilon \qquad (2)$$

where Y is the predicted response, β_0 represents the intercept, β_i, and β_{ij} are the coefficients for linear and interaction terms, respectively, X_i and X_j are the coded predictor variables for the independent factors, and ε is the random error.

Table 1. Independent variables and their coded levels for the optimal design.

Factors	Symbol	Variable level		
		-1	0	+1
Pyrolysis Temperature (°C)	X_1	500	600	700
Pyrolysis Time (min)	X_2	90	135	180
$FeCl_3$ concentrations (M)	X_3	1	2	3

2.4 Adsorption experiment

To study the removal efficiency of Cr(VI) using magnetic biochar derived from rubber seed shells, the magnetic biochar was combined with a Cr(VI) solution at a 30 mg/L concentration. A 0.02 g of the sorbent was added to 20 ml of the Cr(VI) solution. The mixture was placed in a centrifuge tube and shaken at room temperature (25±2 °C) for 24 h at a speed of 150 rpm. Subsequent analyses were performed using an Inductively Coupled Plasma-Optical Emission Spectrometer (ICP-OES) (AVIO 500, Perkin Elmer). The removal efficiency of Cr(VI) can be calculated as Eq. (3).

$$\text{Removal efficiency (R), \%} = \frac{C_i - C_f}{C_i} \times 100 \qquad (3)$$

where C_i and C_f are the initial and final concentrations of Cr(VI), respectively (mg/L).

2.5 Determination of optimal conditions for (Cr(VI)) removal using magnetic biochar from rubber seed shells

This section investigates the identification of optimal conditions and parameters for producing magnetic biochar from rubber seed shells. The primary objective of this optimization was to determine the most effective variable values for the adsorption process. This determination was based on a model derived from experimental data, focusing on maximizing removal efficiency and yield percentage, aiming for the highest removal efficiency and cost-effectiveness. Samples at the optimum conditions were tested in three replicates to verify the accuracy and precision of the data.

3 Results and discussion

3.1 Optimization of Process Parameters for the Preparation of Magnetic Biochar using RSM

Table 2 displays the conditions outlined by the BBD, including actual and coded independent variables. To obtain an optimal fit for the experimental response, linear, quadratic, two-factor interaction, and polynomial equations were employed. Tables 3 and 4 detail the results from the statistical analysis of variance (ANOVA) for the response yield (Y_1) and Cr(VI) removal efficiency (Y_2), respectively. The results of this detailed examination are represented by quadratic equations, as delineated by Eq. (4) for the yield percentage and Eq. (5) for the Cr(VI) removal efficiency.

Table 2. Experimental and predicted yield and Cr(VI) removal efficiency using RSM.

Run	Coded levels			Natural levels			Response	
	X_1	X_2	X_3	X_1	X_2	X_3	Yield (%)	R-Cr(VI) (%)
1	-1	-1	0	500	90	2	52.31	29.42
2	1	-1	0	700	90	2	33.71	36.65
3	-1	1	0	500	180	2	48.78	32.25
4	1	1	0	700	180	2	28.23	28.90
5	-1	0	-1	500	135	1	41.41	27.29
6	1	0	-1	700	135	1	24.58	33.73
7	-1	0	1	500	135	3	57.99	41.78
8	1	0	1	700	135	3	34.59	37.96
9	0	-1	-1	600	90	1	41.58	27.07
10	0	1	-1	600	180	1	31.36	29.72
11	0	-1	1	600	90	3	52.06	41.85
12	0	1	1	600	180	3	46.99	35.91
13	0	0	0	600	135	2	42.28	39.79
14	0	0	0	600	135	2	41.74	37.44
15	0	0	0	600	135	2	40.06	38.74

The experimental results showed that the percentage yield (Y_1) values ranged between 24.58 and 57.99, while the Cr(VI) removal efficiency (Y_2) varied between 27.07 and 41.85. The results are best described by the quadratic regression model, as depicted in Eqs. (4) and (5) for Y_1 and Y_2, respectively. The suitability of the Box-Behnken design (BBD) model was validated through an analysis of variance (ANOVA) at a significance level of $P<0.05$.

$$Y_1 = 41.360 - 9.923X_1 - 3.038X_2 + 6.588X_3 - 1.979X_1^2 + 1.376X_2^2 + 0.261X_3^2 - 0.488X_1X_2 - 1.643X_1X_3 + 1.287X_2X_3 \quad (4)$$

$$Y_2 = 38.657 + 0.810X_1 - 1.029X_2 + 4.961X_3 - 2.646X_1^2 - 4.200X_2^2 - 0.821X_3^2 - 2.639X_1X_2 - 2.566X_1X_3 - 2.147X_2X_3 \quad (5)$$

At a temperature of 500 °C for 135 min and with a $FeCl_3$ concentration of 3 M, the experiment yielded the highest percentage of 57.99%. The results indicated that temperature, time, and $FeCl_3$ concentration significantly impacted the yield percentage, with a significance value of $P<0.05$. The yield percentage decreased as the pyrolysis temperature increased due to the thermal decomposition of cellulose fibers, hemicellulose, and lignin from rubber seed shells [14]. The model demonstrated a high goodness of fit with an R^2 value of 99.38% and an adjusted R^2 value of 98.27%.

As illustrated in Fig.2(a), there is a clear inverse relationship between the yield and two variables: the pyrolysis temperature and the pyrolysis time. Specifically, the yield tends to decrease as the temperature and duration of pyrolysis rise. This could be attributed to the increased thermal breakdown of the materials involved over an extended time and at higher temperatures.

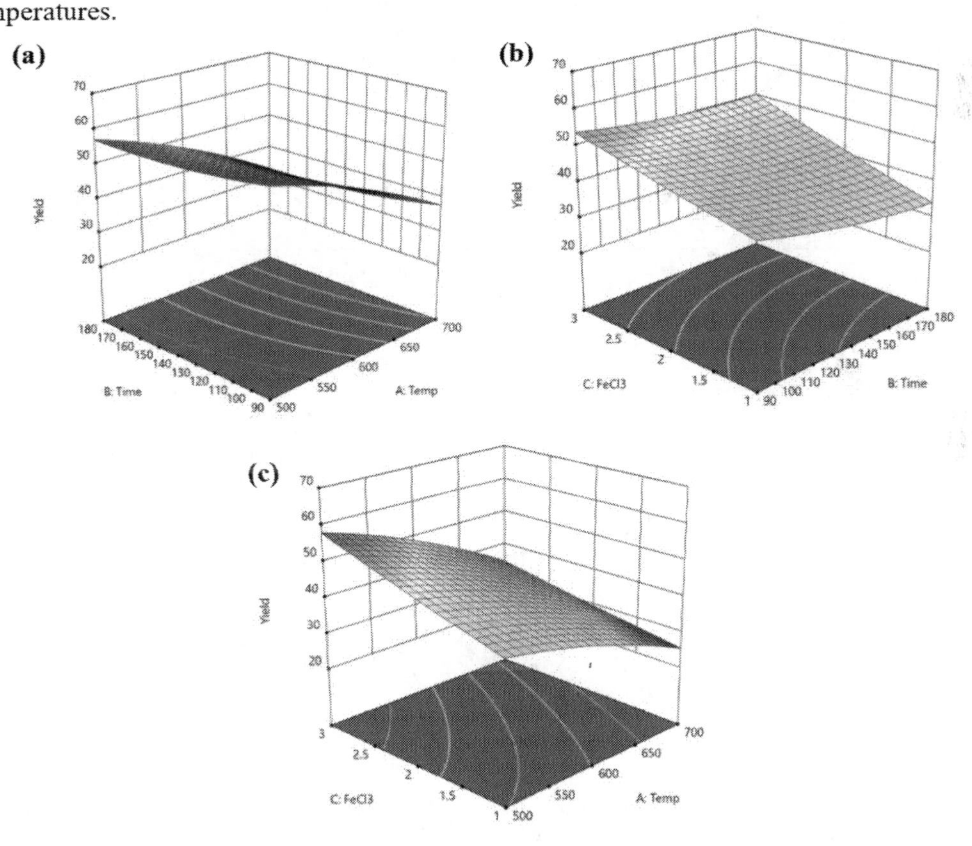

Fig. 2. 3-D surface plots showing the yield based on: (a) the effect of X_1 and X_2, (b) the effect of X_2 and X_3, and (c) the effect of X_1 and X_3.

The concentration of $FeCl_3$ is highlighted as a significant factor influencing yield (Fig.2(b) and (c)). At a concentration of 3 M, $FeCl_3$ significantly increases the yield. However, this positive effect of $FeCl_3$ concentration is mitigated by an increase in pyrolysis temperature and duration. As these two parameters increase, the benefit gained from the $FeCl_3$ concentration starts to decrease, indicating a complex interplay between the chemical concentration and the thermal conditions of pyrolysis.

Table 3. ANOVA and significance test for variables in magnetic biochar yield.

Source	DF	Adj SS	Adj MS	F-Value	P-Value
Model	9	1250.47	138.941	89.23	0.000
Linear	3	1208.62	402.874	258.74	0.000
Temp.	1	787.65	787.648	505.86	0.000*
Time	1	73.81	73.811	47.40	0.001*
$FeCl_3$ conc.	1	347.16	347.161	222.96	0.000*
Square	3	23.48	7.827	5.03	0.057
Temp.*Temp.	1	14.46	14.457	9.28	0.029*
Time*Time	1	6.99	6.993	4.49	0.088
$FeCl_3$ conc.*$FeCl_3$ conc.	1	0.25	0.252	0.16	0.704
2-Way Interaction	3	18.37	6.124	3.93	0.087
Temp.*Time	1	0.95	0.951	0.61	0.470
Temp.*$FeCl_3$ conc.	1	10.79	10.791	6.93	0.046*
Time*$FeCl_3$ conc.	1	6.63	6.631	4.26	0.094
Error	5	7.79	1.557		
Lack-of-Fit	3	5.10	1.701	1.27	0.469
Pure Error	2	2.68	1.340		
Total	14	1258.26			

$R^2 = 0.9938$; adjusted $R^2 = 0.9827$; predicted $R^2 = 0.9303$.
*Significant ($P<0.05$)

Table 4. ANOVA and significance test for variables Cr(VI) removal efficiency using magnetic biochar.

Source	DF	Adj SS	Adj MS	F-Value	P-Value
Model	9	368.860	40.984	52.00	0.000
Linear	3	210.636	70.212	89.08	0.000
Temp.	1	5.245	5.245	6.65	0.049*
Time	1	8.470	8.470	10.75	0.022*
$FeCl_3$ conc.	1	196.921	196.921	249.83	0.000*
Square	3	85.595	28.532	36.20	0.001
Temp.*Temp.	1	25.859	25.859	32.81	0.002*
Time*Time	1	65.136	65.136	82.64	0.000*
$FeCl_3$ conc.*$FeCl_3$ conc.	1	2.486	2.486	3.15	0.136
2-Way Interaction	3	72.629	24.210	30.71	0.001
Temp.*Time	1	27.852	27.852	35.34	0.002*
Temp.*$FeCl_3$ conc.	1	26.337	26.337	33.41	0.002*
Time*$FeCl_3$ conc.	1	18.439	18.439	23.39	0.005*
Error	5	3.941	0.788		
Lack-of-Fit	3	1.173	0.391	0.28	0.838
Pure Error	2	2.768	1.384		
Total	14	372.801			

$R^2 = 0.9894$; adjusted $R^2 = 0.9704$; predicted $R^2 = 0.9330$.
*Significant ($P<0.05$)

There is a significant finding in the relationship between pyrolysis conditions and Cr(VI) removal efficiency, as shown in Fig. 3(a). Increasing pyrolysis temperature and time improves Cr(VI) removal efficiency. In Figs. 3(b) and (c), the role of $FeCl_3$ concentration in this process is explored. It is important to note that the concentration of $FeCl_3$ is not a passive variable; rather, it plays a crucial role in determining the efficiency of Cr(VI) removal. An increased concentration of $FeCl_3$ may lead to the presence of additional iron and aromatic functional groups [15]. This highlights the significance of carefully considering the concentration of $FeCl_3$ when assessing and optimizing the removal process. Higher concentrations of $FeCl_3$ could potentially enhance the interaction between the reactive species, leading to more effective sequestration of Cr(VI). This observation underscores the importance of optimizing the pyrolysis conditions and the chemical additives, such as $FeCl_3$, to achieve maximum removal efficiency for Cr(VI).

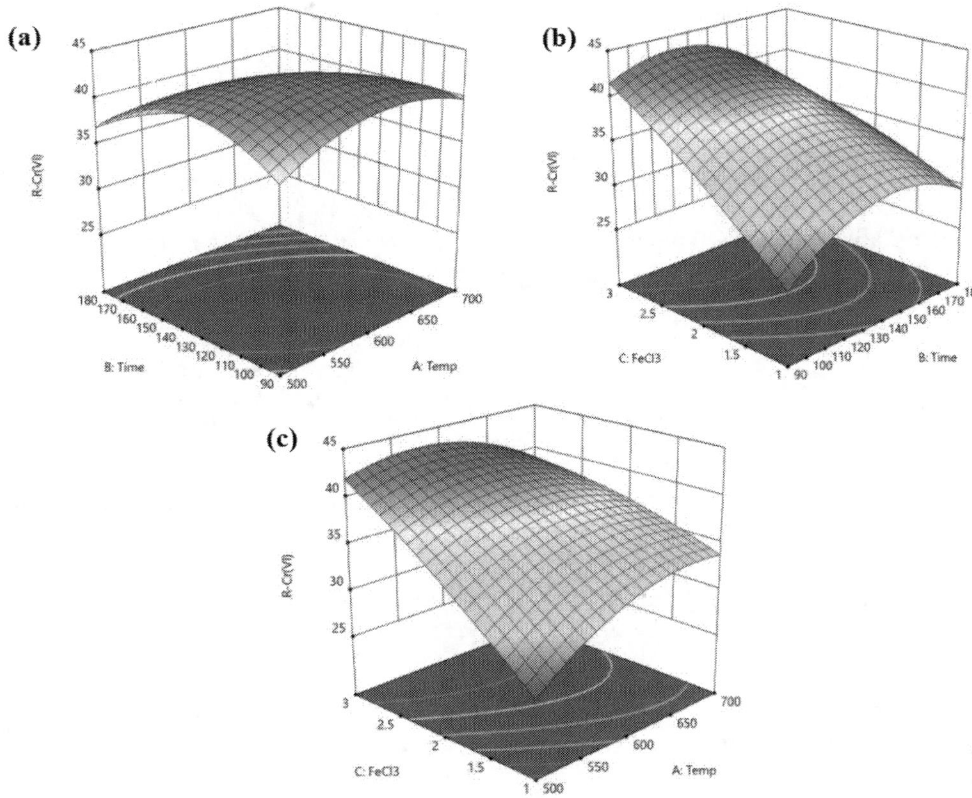

Fig. 3. 3-D surface plots showing the Cr(VI) removal efficiency based on: (a) the effect of X_1 and X_2, (b) the effect of X_2 and X_3, and (c) the effect of X_1 and X_3.

3.2 Optimization and validation

In optimizing the production of magnetic biochar, a balance had to be maintained between the yield percentage and Cr(VI) removal efficiency. While aiming for the highest yield possible might appear beneficial, experiments indicated that maximizing yield led to a decrease in Cr(VI) removal efficiency. This finding influenced the decision to target a yield of around 50%. This specific yield limit was chosen because it provided a satisfactory amount of magnetic biochar and, remarkably, ensured a high Cr(VI) removal efficiency. Balancing quantity and quality was essential for achieving efficient Cr(VI) removal. As shown in Table 5, under optimal conditions of a pyrolysis temperature of 580 °C, a pyrolysis time of 130 min, and a $FeCl_3$ concentration of 3 M, the yield and Cr(VI) removal efficiency closely matched model predictions.

Table 5. Confirmation of the experimental results from optimized conditions.

	Temperature (°C)	Time (min)	$FeCl_3$ concentration (M)	Yield (%)	R-Cr(VI) (%)
Predicted value	577.30	132.39	2.89	50.00	42.73
Experimental value	580.00	130.00	3.00	48.63 ±1.53	41.29 ±1.05

4 Conclusions

The production of magnetic biochar from rubber seed shells and its effectiveness for Cr(VI) removal was investigated. Utilizing the Box-Behnken Design, optimal conditions were identified that ensure a balance between yield percentage and Cr(VI) removal efficiency. Specifically, a yield threshold of approximately 50% was targeted, demonstrating high Cr(VI) removal efficiency. Optimization of these process parameters suggested a pyrolysis temperature of 580 °C, a pyrolysis time of 130 min, and a $FeCl_3$ concentration of 3 M. The observed yield and Cr(VI) removal efficiency of 48.63% and 41.29%, respectively, closely matched with predictions under these optimum conditions. These findings are significant as a sustainable method for preparing magnetic biochar, displaying its potential as an effective sorbent. Furthermore, by optimizing yield and removal efficiency, the process maximizes resource utilization and enhances the overall efficiency of Cr(VI) removal from solutions. This study provides the way for further research in waste-to-value conversion. Additionally, it emphasizes the significance of magnetic biochar in addressing environmental issues. Future work could explore increasing the amount of magnetic biochar to further enhance Cr(VI) removal efficiency.

Acknowledgements

This work was supported by King Mongkut's Institute of Technology Ladkrabang (KMITL) and the NSRF (FRB660065/0258-RE-KRIS/FF66/33).

References

1. T. Q. Nguyen, V. Sesin, A. Kisiala and R. J. N. Emery: Phytohormonal Roles in Plant Responses to Heavy Metal Stress: Implications for Using Macrophytes in Phytoremediation of Aquatic Ecosystems. Environ Toxicol Chem, **40**(1), 7-22 (2021)
2. L. Yan, L. Kong, Z. Qu, L. Li and G. Shen: Magnetic Biochar Decorated with ZnS Nanocrytals for Pb (II) Removal. ACS Sustain Chem Eng, **3**(1), 125-132 (2014)
3. M. Balali-Mood, K. Naseri, Z. Tahergorabi, M. R. Khazdair and M. Sadeghi: Toxic Mechanisms of Five Heavy Metals: Mercury, Lead, Chromium, Cadmium, and Arsenic. Front Pharmacol, **12**, 643972 (2021)
4. M. M. Uddin, M. C. M. Zakeel, J. S. Zavahir, F. Marikar and I. Jahan: Heavy Metal Accumulation in Rice and Aquatic Plants Used as Human Food: A General Review. Toxics, **9**(12) (2021)
5. Oae: Rubber Plantations. Office of Agricultural Economics, Ministry of Agriculture and Cooperatives, BANGKOK (2022)

6. W. Roschat, T. Siritanon, B. Yoosuk, T. Sudyoadsuk and V. Promarak: Rubber seed oil as potential non-edible feedstock for biodiesel production using heterogeneous catalyst in Thailand. Renew Energ, **101**, 937-944 (2017)
7. S. Chaireh, K. M. Szécsényi, P. Boonsuk and K. Kaewtatip: Preparation of rubber seed shell powder by planetary ball milling and its influence on the properties of starch foam. Ind Crops Prod, **135**, 130-137 (2019)
8. X. Li, C. Wang, J. Zhang, J. Liu, B. Liu and G. Chen: Preparation and application of magnetic biochar in water treatment: A critical review. Sci Total Environ, **711**, 134847 (2020)
9. X. Zhang, P. Zhang, X. Yuan, Y. Li and L. Han: Effect of pyrolysis temperature and correlation analysis on the yield and physicochemical properties of crop residue biochar. Bioresour Technol, **296** (2020)
10. Y. Yu, Q. An, L. Jin, N. Luo, Z. Li and J. Jiang: Unraveling sorption of Cr (VI) from aqueous solution by FeCl(3) and ZnCl(2)-modified corn stalks biochar: Implicit mechanism and application. Bioresour Technol, **297**, 122466 (2020)
11. Y. Wang, Q. Yang, J. Chen, J. Yang, Y. Zhang, Y. Chen, X. Li, W. Du, A. Liang, S. H. Ho ,and J. S. Chang: Adsorption behavior of Cr(VI) by magnetically modified Enteromorpha prolifera based biochar and the toxicity analysis. J Hazard Mater, **395**, 122658 (2020)
12. L. Xiao, H. Lu, J. Li, Q. Kong, Y. Lan and D. Wang: Preparation of biochar from constructed wetland plant and its adsorption performance towards Cu2+. Environ Sci Pollut Res (2022)
13. A. a. H. Saeed, N. Y. Harun, S. Sufian, M. R. Bilad, B. A. Nufida, N. M. Ismail, Z. Y. Zakaria, A. H. Jagaba, A. a. S. Ghaleb and B. N. S. Al-Dhawi: Modeling and Optimization of Biochar Based Adsorbent Derived from Kenaf Using Response Surface Methodology on Adsorption of Cd2+. Water, **13**(7) (2021)
14. X. Zhang, P. Zhang, X. Yuan, Y. Li and L. Han: Effect of pyrolysis temperature and correlation analysis on the yield and physicochemical properties of crop residue biochar. Bioresour Technol, **296**, 122318 (2020)
15. Y. Zhou, G. Liu, J. Liu, Y. Xiao, T. Wang and Y. Xue: Magnetic biochar prepared by electromagnetic induction pyrolysis of cellulose: Biochar characterization, mechanism of magnetization and adsorption removal of chromium (VI) from aqueous solution. Bioresour Technol, **337**, 125429 (2021)

Production of bio-based paper coating made with the addition of eggshell powder

Zsófia Kóczán[1*], and *Zoltán* Pásztory[1]

[1] University of Sopron, 4 Bajcsy Zs. str. Sopron, 9400 Hungary

Abstract. A chitosan based coating was prepared where eggshell powder was added. Eggshell is a product that appears in large numbers in everyday life, so it has great potential as waste, which can be used to turn it into a valuable material. Cellulose sheets were made from secondary fiber, and after coated with the chitosan based solution to investigate the properties of the sheets. The results of the treated handsheets were compared with the results of the control sheet. The control sheets were compared with 4 different type of paper: coatings of chitosan with 5wt% and 10wt% eggshell powder, only chitosan coating, and paper where eggshell powder was added to the paper pulp. The use of the eggshell powder as an additive in the coating would be a promising material, because it can improve in a large extent the water resistance and the air barrier properties of the paper.

1 Introduction

The raw material of paper is cellulose, which is the most common natural polymer on earth. Cellulose is a renewable material, the sources of which can be the wood, annual plants, marine animals, algae, fungi [1]. In other words, sustainable raw materials are used for the production of paper products. Paper and cardboard products are biodegradable, thus not polluting the environment, but it can also be reused, i.e. a product that fits into the model of the circular economy, which is mostly used in the packaging sector as an environmental friendly alternative. In addition to these favorable properties, fiber-based products have weak barrier properties against water, oil and oxygen. Polymers, aluminum foils, and waxes are used to create coatings on the paper, which can be used to improve poor barrier properties [2-5]. These coatings often adversely affect recyclability and biodegradability because they are fossil-based, synthetic materials [6].

In recent decades, sustainable biopolymers have been extensively researched as an alternative to fossil-based plastics. Among the biopolymers, polysaccharides, proteins, lipids and polyesters can also be used to produce films and coatings [7]. A lot of emphasis was placed on chitosan research because of its favorable properties: excellent film-forming properties, non-toxic, biodegradable, antibacterial and it's naturally occuring [8, 9]. The production of chitosan requires chitin, which can be found in shrimps, crabs and the shells of insects. The waste product of the seafood industry can be used to produce chitosan. Chitosan is produced by deacetylating chitin with sodium hydroxide [8, 10]. Chitosan is semi-

* Corresponding author: koczan.zsofia@uni-sopron.hu

crystalline in the solid phase, and its dissolution requires a dilute organic acid medium [11]. The barrier properties such as oxygen and water vapour permeability and mechanical properties of chitosan are weak [10, 12].

Eggshell is a by-product that appears in the food processing and manufacturing sectors in large numbers in everyday life. Since eggs are often used in households, restaurants, or industry. The eggshell and its membrane are largely composed of inorganic components and contain only a small proportion of organic components. 95% is calcium carbonate, while 5% contains collagen, sulfated polysaccharides and other proteins [13, 14]. Between 2019 and 2021, egg production in Hungary averaged 2,550 million eggs based on statistics [15]. Most of the eggshell waste ends up in landfills without pretreatment [16]. Eggshells are non-hazardous waste in the environment, however, the organic protein matrix in them attracts rats and worms, which can lead to public health problems [17]. In the past decade, many studies have focused on how eggshells or eggshell membrane can be used as a value-added product in different fields e.g. ceramic wall tile, alginate, cornstarch and chitosan based biodegradable or edible film [14, 17-20].

The aim of the research is to create a coating for paper industry products that does not prevent the biological degradation of the paper. For this a chitosan-based coating was prepared. By adding ground eggshell to chitosan, we investigated whether it improves the weak mechanical and barrier properties of chitosan. We also investigated how much the paper's properties are affected by the applied coating. Although chitosan films and coatings are examined in many researches, its composites with eggshell as a paper coating material are not included in the reviewed publications.

2 Materials and methods

2.1 Materials

The eggshells were collected from local households in Sopron, Hungary. Chitosan (medium molecular weight) was procured from Sigma-Aldrich. Acetic acid was purchased from Molar Chemicals Ltd., Hungary. Deionized water was used for making the solution.

2.2 Preparation of eggshell powder

After collection, the eggshells were washed several times with tap water. Then with deionized water to remove impurities. After that, the eggshells were dried for 8 hours at 50°C in a hot air oven. The eggshell was blended into fine particles using a Planetary Ball Mills during 2 minutes and 400 rpm with 10 mm diameter balls (Retsch, PM 100).

2.3 Preparation of coating material

Chitosan solution of 1 wt% was prepared by dissolving chitosan powder in 1 mg/mL acetic acid solution for 24 h on a magnetic stirrer at room temperature (21 ± 2 °C). Then 5 wt% and 10 wt% of eggshell powder was added to the chitosan solution and gently stirring for 20 h. Later it was kept in a closed container in the fridge.

2.4 Preparation of samples

To prepare the cellulose sheets, secondary fiber was used and prepared 0.6 wt% suspension in a disintegrator with tap water. Sheet formation was performed on a Rapid-Köthen sheet former (Ernst Haage, D-45476). The wet sheets were dried in a sheet dryer under vacuum at

90°C for 3 minutes. 5 different type of paper was made: control paper which contains only secondary fiber (C) paper with coating of chitosan with 5 wt% eggshell powder (E1), paper with coating of chitosan with 10 wt% eggshell powder (E2), paper which only coated with chitosan (CH), and paper where eggshell powder (10 wt% for dry basis of fiber) was added to the paper pulp (PEGS).

For coating a brush was used where 4 g of coating was applied to the surface. Then it was dried on a Rapid-Köthen sheet former's dryer under vacuum at 90°C for 1 minutes. The finished sheets were kept in a sealed bag at room temperature, protected from light, until the tests started.

2.5 Conditioning

Before the measurements, the sheets were conditioned according to MSZ EN 20187 at 23°C and 50% relative humidity. The tests were performed in the same environment.

2.6 Thickness, grammage

The thickness of the sheets was measured according to MSZ ISO 534 with a digital micrometer (Lorentzen & Wettre) to an accuracy of 100 μm. The grammage was determined according to MSZ ISO 536. After weighing the mass (0.0001 g) on an analytical balance (Sartorius), the grammage was expressed in g/m^2. The values which were given are the averages with the standard deviations.

2.7 Examination of paper properties

The air permeability test was performed according to ISO 5636-5 using a Gurley densometer (Lorentzen & Wettre). The tensile test was performed according to MSZ ISO 1924 standard on an Alwetron TH1 equipment (Lorentzen & Wettre). For the test the samples were cut out of the sheet in the size of 15 x 100 mm). The values which given are averages, with standard deviations, according to standard measurements.

2.8 Water resistance

To study the water resistance of the samples a Cobb test and a microscopy test were performed. The test was performed during 30 seconds, where 20 seconds was the wetting process and 10 seconds was the drying process where samples was put between blotting paper. The absorption was expressed in g/m^2. The microscopics images was taken with a Tagarno FHD Prestige digital microscopy, where a water drop dropped at the surface of the sample then images was taken immediately at the dropping time and after one minute, in that cases where were sense to take images after one minute.

2.9 Scanning electron microscopy (SEM)

The SEM images of the samples and the coated surface were taken with a Hitachi S-3400 N scanning electron microscope (Tokyo, Japan) with an accelerating voltage of 10 kV. The magnification was 100 and 300.

2.10 Fourier-transform infrared spectroscopy (FTIR)

The FTIR spectra were collected with a Jasco FT/IR 6300 spectrophotometer (Tokyo, Japan) equipped with an ATR PRO 470-H. The spectra were recorded in the range of 4000 to 1000 cm^{-1} in the transmission mode with 40 scans per samples.

2.11 Statistical analysis

The one-way analysis of variance (ANOVA) and Tukey's multiple range test were used to compare the differences between the mean values of the tested samples. Differences below the significance level of 0.05 were considered statistically significant.

3 Results and discussion

3.1 Thickness, grammage

The mean values and standard deviations of thickness and grammage are shown in Table 1. It shows clearly that the control sheet is the thinner, and the treated sheets are thicker, because it contains an extra layer on the surface or extra additive in the structure. The treated sheets showed significant difference in all cases compared to the control sheet. However, compared the treated sheets to each other only between the E1 and E2 and between the E2 and CH there are significant difference.

Table 1. Thickness and grammage of the tested sheets.

	Control	E1	E2	CH	PEGS
Thickness (μm)	195±4 ^{a}	213±7 ^{b}	223±8 ^{c}	215±7 $^{b\,c}$	207±6 ^{b}
Grammage (g/m^2)	88±4 ^{a}	97±3 ^{b}	103±4 $^{b\,c}$	86±4 $^{a\,d}$	91±7 $^{a\,e}$

Values with different letters in the same columns indicate significant differences ($p < 0.05$).

3.2 Air permeability and tensile test

Table 2. shows the results of the air permeability test of the samples, where there is no significant difference between the control, the E1, the chitosan coated and the eggshell powder contained paper. E1 and E2 shows better airtight properties, however only the E2 results shows significant difference compared to the other results. The time required to penetrate 100 cm^3 of air increased by 27 times more at E1, and by 175 times more at E2 compared to the control sheets. In case of E2 outstanding improvement can be observed. Consequently, we can say only the chitosan coating cannot improve the air barrier properties, but when eggshell powder was added to the chitosan it can improve the air barrier properties. In addition, by increasing the eggshell powder, the time required for penetration also increases.

Table 2. Air permeability result of the samples.

	Control	E1	E2	CH	PEGS
Air permeability (Gurley s)	0.9±0.1 ^{a}	24.7±14.2 ^{a}	157.3±40.8 ^{b}	1.6±0.3 ^{a}	1.09±0.2 ^{a}

Values with different letters in the same columns indicate significant difference ($p < 0.05$).

The results of the tensile test, so the tensile strength and the strain of the samples are shown in the Table 3. The best results in tensile strength was achieved with the pure chitosan coating. The addition of the eggshell powder decreased the strength of the paper, and it weakens further by increasing the amount of eggshell powder. However, the addition of the eggshell powder in 5 wt% was increased the strain. Except for the PEGS sheet, all results proved significantly different compared to the control sheet in the results of the tensile test and also in the strain results.

Table 3. Results of the tensile test.

	Control	E1	E2	CH	PEGS
Tensile strength (kN/m)	0.8±0.1 [a]	1.6±0.2 [b]	1.2±0.1 [c]	1.7±0.1 [b]	0.8±0.3 [a]
Strain (%)	0.9±0.2 [a]	2.0±0.3 [b]	1.3±0.3 [c]	1.7±0.3 [d]	1.1±0.2 [a c]

Values with different letters in the same columns indicate significant difference ($p < 0.05$).

3.3 Water resistance

3.3.1 Cobb test

The results of the water absorption test shows huge differences. Figure 1. shows the control (a), the E1 (b), E2 (c) and the CH (d) sheets after wetting. No picture from PEGS sheet because it shows totally the same than the control sheet. On the control sheet (Fig. 1. a) we can observe that almost the whole paper was wet, only some lighter parts can see which are the dry parts. The E1 sheet (Fig.1. b) shows better result, since the water surface is almost only visible in a circular shape (which is the shape of the Cobb tester, and on this circular surface water directly touches the paper). This means that the fibers in contact with water did not absorb as much water as in the case of the control, since the water did not spread over a larger surface than the surface in contact with the Cobb tester. In case of E2 sheet (Fig 1. c) we can see the best results, because even on the surface in contact with water, you can see lighter spots, which means that it has not absorbed as much water as the area around it. CH sheet (Fig. 1. d) shows better results than control, but it was able to absorb more water compared to eggshell powder-containing coatings. Therefore, the addition of the eggshell powder to the chitosan can improve the water resistance of the paper.

The mean values and standard deviations of the Cobb test are shown in Table 4. The results of the Cobb test shows that E1 and E2 sheets showed significant improvement compared to the control sheet, which we can clearly observe with the images. E1 and E2 compared to the PEGS sheet also show significant improvement, but other cases the statistical analysis did not show significant differences between the results.

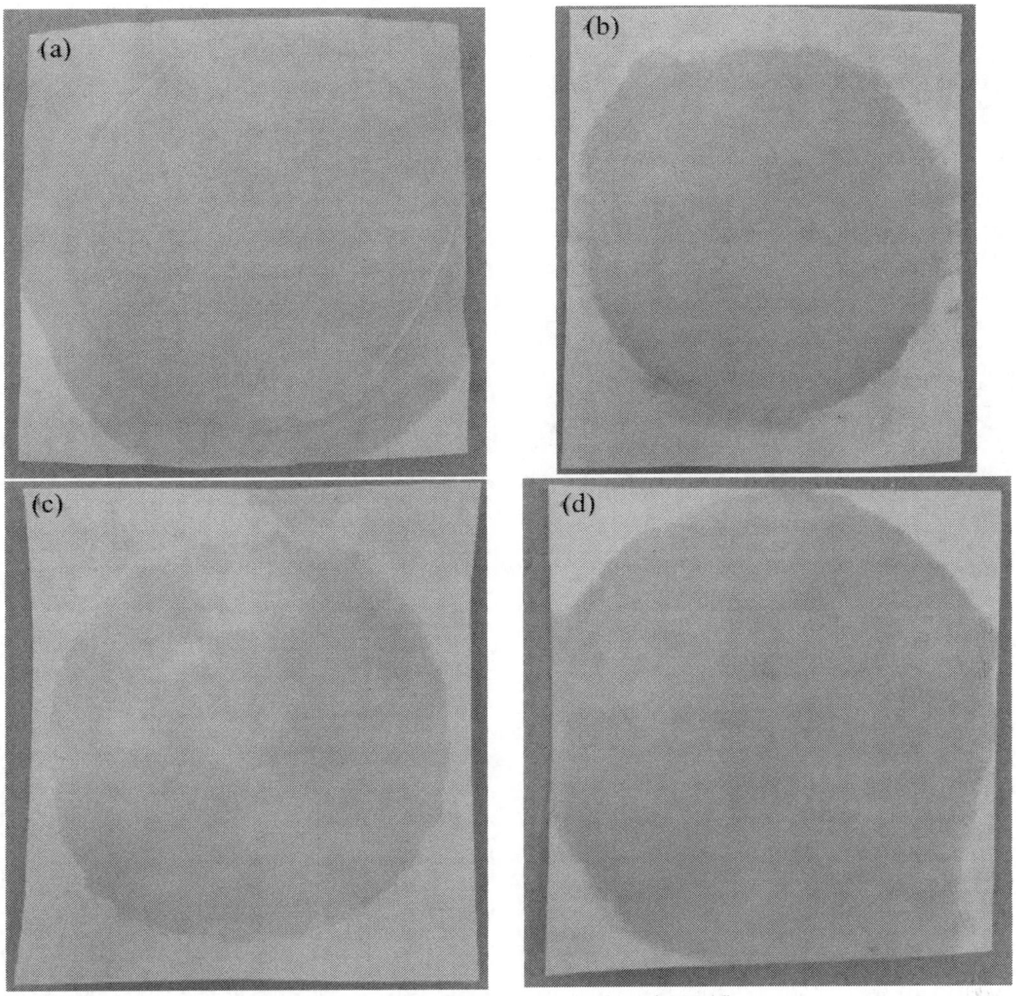

Fig. 1. The water absorption test of control (a); E1 (b); E2 (c) and CH (d).

Table 4. Results of the Cobb test.

	Control	E1	E2	CH	PEGS
Absorbed water (g/m^2)	284.9±7.9 ^{a}	159.6±22.2 ^{b}	137.6±36.3 $^{b\,c}$	208.6±21.2 $^{a\,b}$	247.5±31.9 ^{a}

3.3.2 *Microscopic images*

Figure 2. (a) shows the control paper where the dropped water drop spread over the surface of the paper, and the paper immediately absorbed the water. E1 sample (Fig. 2. b,c) and CH samples (Fig. 2. f, g) showed similar results, which were better results compared to the control sheet. After the drop, the water drop can be seen on the surface of the paper and not spreads out. Even after 1 minute, the water drop is still clearly visible on the surface, but at this point a darker spot can be seen around it, which means that absorption has begun, but much slower than with the control paper. E2 samples showed the best results from all the samples. After 1 minute the drop spread in a small amount compared to the initial state, but darker parts around the water drop cannot be visible, which means the fibers did not absorb much water. PEGS (Fig. 2. h) showed that the eggshell powder as an additive of the paper pulp cannot improve the water resistance properties of the paper, because it showed the same results, then the control sheet.

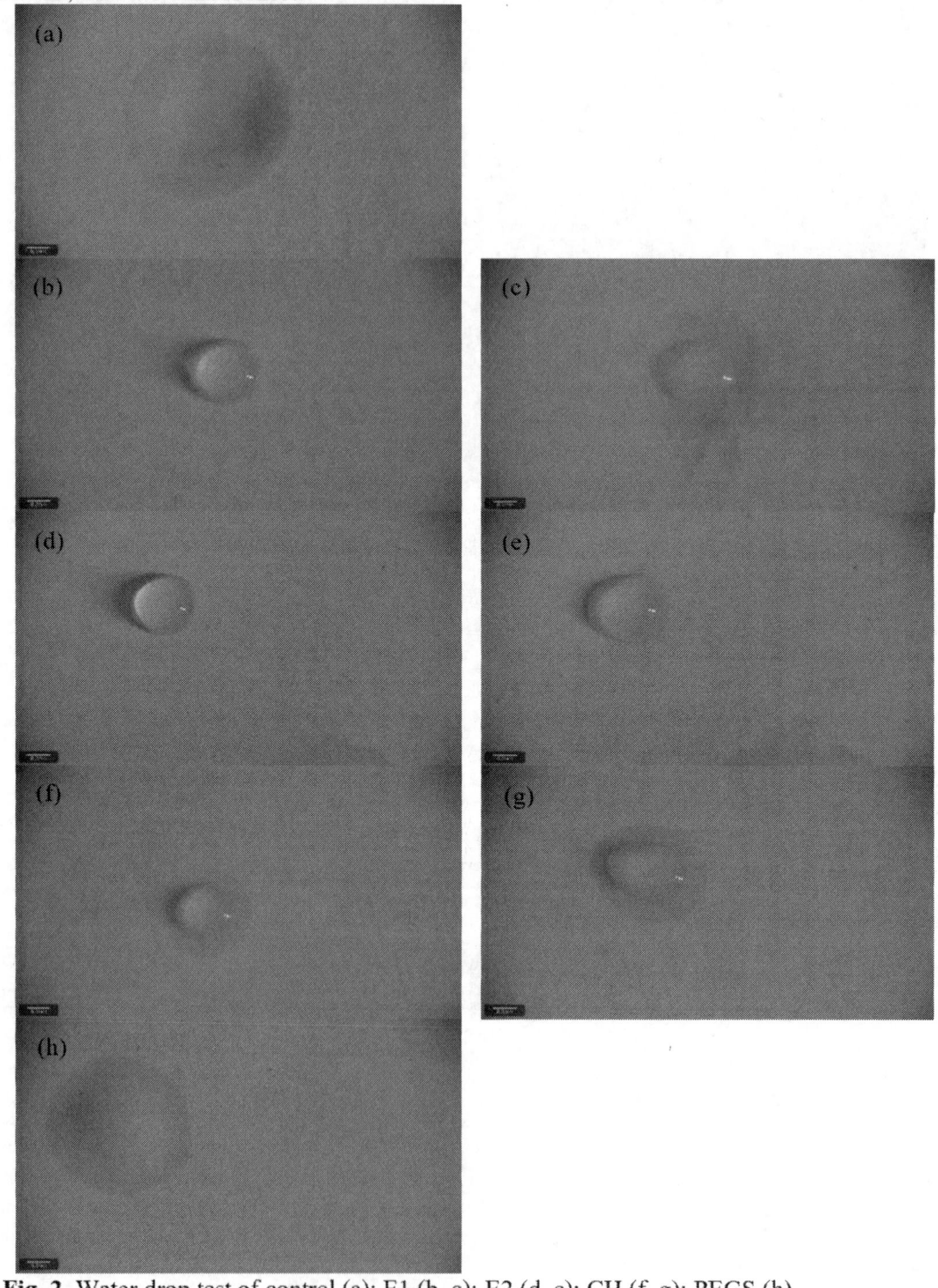

Fig. 2. Water drop test of control (a); E1 (b, c); E2 (d, e); CH (f, g); PEGS (h).

3.4 SEM analysis

The surface structure of the paper and the coating was studied using a scanning electron microscopy. Figure 3 shows the SEM images with magnification of 100× and 300×. In the

 https://doi.org/10.1051/e3sconf/202451402003

control picture (Fig. 3. a, b) only the fibers can be observed. These are long fibers, where the length at least 0.5 mm, but there are some fibrils, which are thinner broken fibers.

Picture of the E1 (Fig. 3. c, d) clearly shows the eggshell parts on the surface of the paper. In 300× magnification we can see obscure parts which can be the small parts of the protein-rich membrane layer. In the picture of E2 (Fig. 3. e, f) can be observed much more eggshell parts because of this coating contained the eggshell particles in the largest proportion. The coating covers the surface well, however we can see parts where is no coverage, so we can see the fibers. It can be concluded that the coating process was not totally perfect. The chitosan coating (Fig 3. g, h) shows nothing visible. It is quite similar to the control paper. The Figure 3. (i, j) shows the PEGS sheet. In the 100× magnification a large white spot is observed, which is believed to be clumped eggshell powder that has not been properly mixed into the pulp. Since the eggshell was added at the beginning of the papermaking process, it can be assumed that a part of the eggshell powder was removed along with the water during the dewatering step. That is why the eggshell parts only can be observed in a small extent.

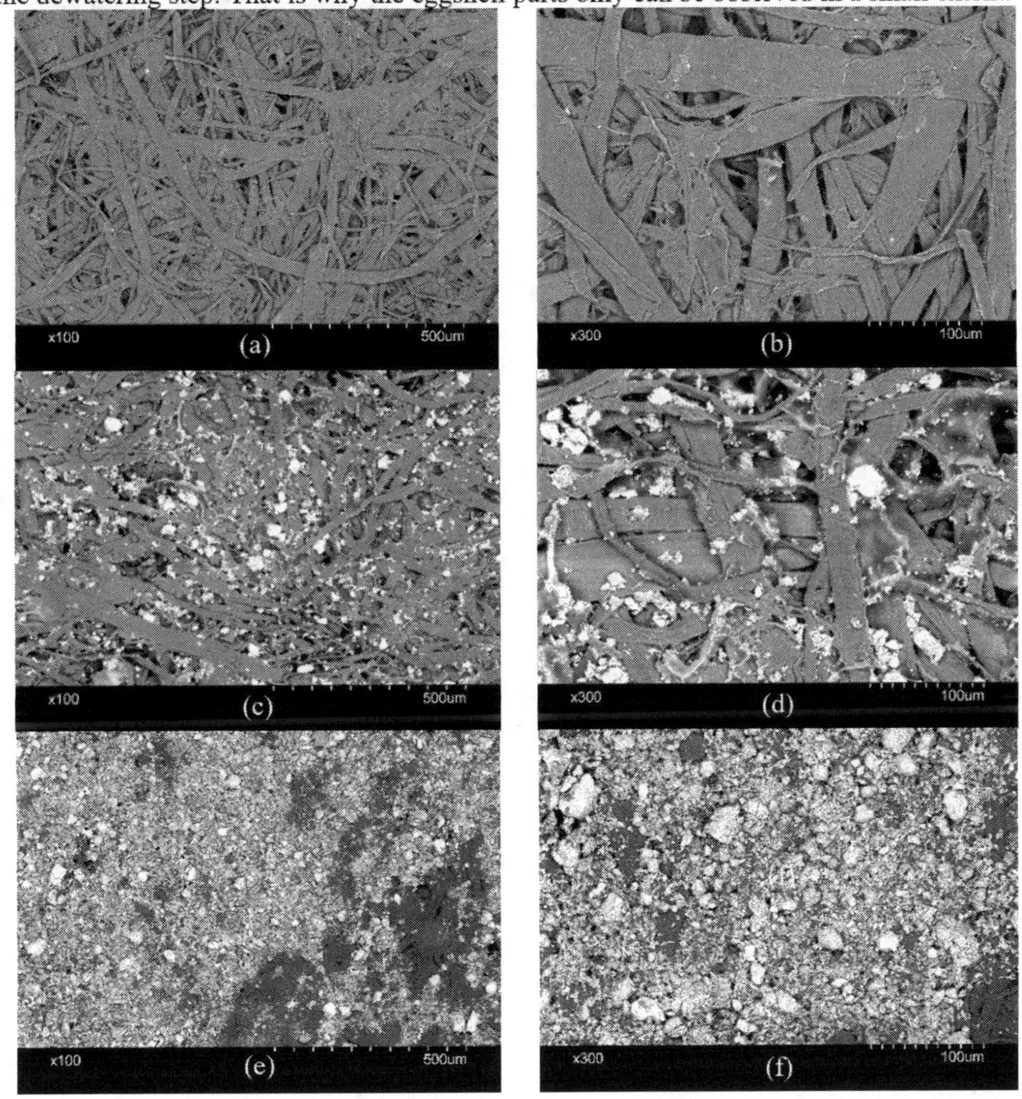

https://doi.org/10.1051/e3sconf/202451402003

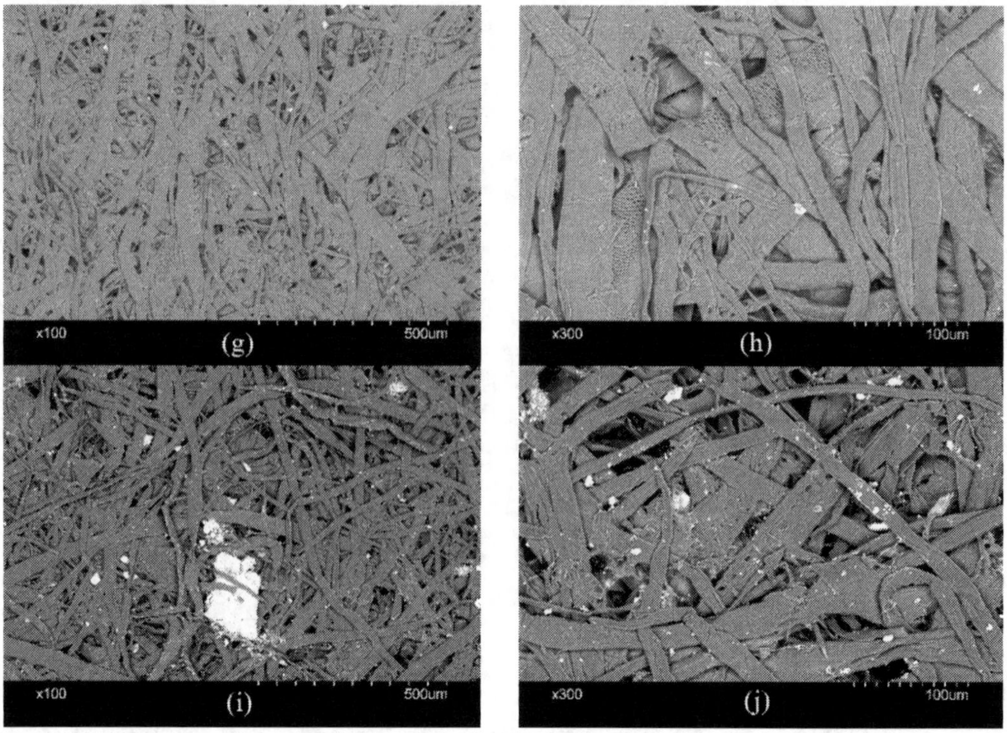

Fig. 3. SEM images of control (a,b); E1 (c,d); E2 (e,f); CH (g,h); PEGS (i,j).

3.5 FTIR analysis

With the use of infrared spectroscopy the intermolecular interaction of the control and treated samples were examined. Figure 4. showed the FTIR spectra of the samples. At 3330 cm^{-1} a broad and strong band can be observed with all samples which is assigned to the O—H stretching vibration of H-bonded hydroxyl groups. The stretching vibration of C—H can be observed at the adsorption of 2896 cm^{-1}. The peak at 1622 cm^{-1} is due to the vibration of the adsorbed H_2O molecules. C—H bending of CH_2 can be seen at 1427 and 1315 cm^{-1}. The peak at 1368 cm^{-1} is the C—H bending vibration of CH. The band at 1160 cm^{-1} is due to the C—O—C stretching vibrations of β-(1,4)-glycosidic bonds. The in-plane ring C—O—C stretching vibration can be observed at 1160 cm^{-1}. Two strong peaks can be observed at 1053 and 1029 cm^{-1}correspond to the C—O stretching vibration at CH—OH and CH2—OH [21-24].

The spectra shapes are similar in all cases, however there are some differences in the peak positions of the different samples. In one case, we can observe a different peak compared to the control and PEGS sheet. Around 1550 cm^{-1} there is a peak in the case of the chitosan contained samples, which can be the N—H [25].

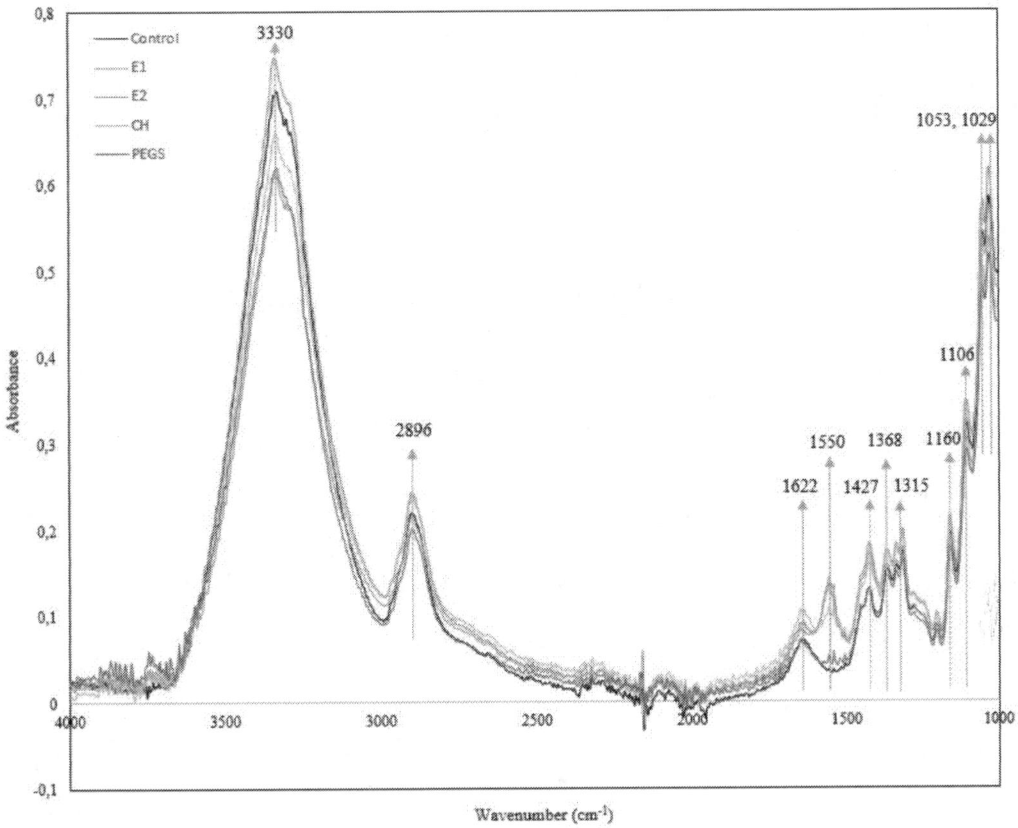

Fig. 4. Absorption spectra of the samples.

4 Conclusion

The results of the research showed that the air barrier property of the secondary fiber sheet can be improved to be coated with the eggshell powder contained chitosan film. However, the paper which contained in the chitosan coating 5 wt% eggshell powder showed an improvement compared to the control film but it was not significant. In the case of 10 wt% eggshell powder the improvement was significant compared to the other tested samples. According to the results of the tensile test the E1, E2 and CH sheets showed improvement compared to the control and PEGS sheets. The best outcomes showed by the E1 and CH sheets. The Cobb test showed that the water resistance of all the treated paper improved compared to the control sheet. In the water resistance, the significant improvement was achieved in the case of E1 and E2 sheets compared to the control and PEGS sheets. During the water drop test, the microscopic pictures showed that the highest improvement was achieved the E2 sheet, after followed the E1 and CH, which showed similar results. However, the PEGS sheet showed totally the same than the control sheet. The SEM images clearly showed that the most closed surface was achieved with the 10 wt% eggshell powder contained chitosan solution, which is explain the improved airtight and water resistance properties. The absorption spectra of the FTIR were similar in all cases, we can observe some differences only in the peak positions of the different samples. It can be concluded that the eggshell contained chitosan coating can improve the air barrier and water resistance properties of the paper compared to the chitosan-coated paper. However, the addition of the eggshell powder was not able to improve the tensile properties of the samples.

This article was made in frame of the project TKP2021-NKTA-43 which has been implemented with the support provided by the Ministry of Innovation and Technology of Hungary (successor: Ministry of Culture and Innovation of Hungary) from the National Research, Development and Innovation Fund, financed under the TKP2021-NKTA funding scheme.

References

1. L. Brinchi, F. Cotana, E. Fortunati, J.M. Kenny, Carbohydr. Polym. **94**, 154–169 (2013)
2. P.K. Kunam, D. Ramakanth, A. Konala, K. Gaikwad, Biomass Convers. Biorefin. 1-16 (2022)
3. A. Adibi, D. Valdesueiro, J. Mok, N. Behabtu, C. Lenges, L. Simon, T.H. Mekonnen, Carbohydr. Polym. **282,** 119121 (2022)
4. L. Dai, X. Xi, X. Li, Y. Du, Y. Lv, W. Wang, Y. Ni, Carbohydr. Polym. **271**, 118425 (2021)
5. D. Raheem, Emir. J. Food Agric. **25,** 177-188 (2012)
6. K. Khwaldia, E. Arab-Tehrany, S. Desobry, Compr. Rev. Food Sci. Food Saf. **9**, 82–91 (2010)
7. V. Rastogi, P. Samyn, Coatings **5**, 887–930 (2015)
8. A. Jiang, R. Patel, B. Padhan, S. Palimkar, P. Galgali, A. Adhikari, I. Varga, M. Patel, Polymers **15**, 2235 (2023)
9. H. Singhi, L. Kumar, P. Sarkar, K. Gaikwad, J. Food Meas. Charact. **17,** (2023)
10. T. Liu, J. Li, Q. Tang, P. Qiu, D. Gou, J. Zhao, Foods **11,** 1490 (2022)
11. M. Flórez, E. Guerra-Rodríguez, P. Cazón, M. Vázquez, Food Hydrocoll. **124**, 107328 (2022)
12. D. Hu, X. Liu, Y. Qin, J. Yan, J. Li, Q. Yang, Food Qual. Saf. **6**, (2022)
13. S.C. Onwubu, V. Anisa, S. Shalini, M.K. Krishnan, J. Appl. Biomater. Funct. Mater. **15**, 341-346 (2017)
14. V.M. Joseph, R. Kobun, R.A. Azhar, N. Huda, E.H. Kana, W.X.L. Felicia, M.N. Nur'Aqilah, N.F.A. Halid, J. Funct. Biomater. **13,** 67 (2022)
15. https://www.ksh.hu/stadat_files/mez/hu/mez0034.html [Viewed: 10.08.2023]
16. Y. Gao, C. Xu, Catal. Today **190,** 107–111 (2012)
17. M.N. Freire, J.N.F. Holanda, Ceramica **52**, 240–244 (2006)
18. R. Mohammadi, M.A. Mohammadifar, M. Rouhi, M. Kariminejad, A.M. Mortazavian, E. Sadeghi, S. Hasanvand,. Int. J. Biol. Macromol. **107**, 406–412 (2018)
19. B.C.W. Webb, S. Rafferty, A.J. Vreugdenhil, Polymers **14**, 383 (2022)
20. V. Villanueva, F. Valdes, R.N. Zuniga, M.G. Villamizar-Sarmiento, E. Soto-Bustamante, P. Romero-Hasler, A.L. Riveros, J. Tapia, J. Lisoni, F. Oyarzun-Ampuero, C. Valenzuela, Food Hydrocoll. **142** (2023)
21. Y. Yan, C. Wen, M. Jin, L. Duan, R. Zhang, C. Luo, J. Xiao, Z. Ye, B. Gao, P. Liu, Y. Tang, Chem. Res. Chin. Univ. **35**, 586–591 (2019)
22. Y. Ma, J. Wang, Y. Zhang, J. Therm. Anal. Calorim. **129**, 1225–1232 (2017)

23. X. Colom, F. Carrillo, F. Nogués, P. Garriga, Polym. Degrad. Stab. **80**, 543–549 (2003)
24. N. Abidi, L. Cabrales, C.H. Haigler, Carbohydr. Polym. **100**, 9–16. (2014)
25. J. Novaes, E.A.S. Filho, P.M.F. Bernardo, E.R. Yapuchura, Polímeros **30**, e2020015 (2020)

Feasibility Study of Using Wave-PV-Wind Hybrid System for AL Hallaniyat Island

Abdullah Al-Badi[1]*, *Abdulmajeed* Al Wahaibi[2], and *Humaid* Al-Badi[3]

[1]Department of Electrical & Computer Engineering, College of Engineering, Sultan Qaboos University, Muscat, Oman; email: albadi@squ.edu.om
[2]Department of Petroleum & Natural Gas Engineering, College of Engineering, Sultan Qaboos University, Muscat, Oman; email: Majeed.al.Wahaibi20@alumni.colostate.edu
[3]Civil Aviation Authority, Directorate General of Meteorology, Oman

Abstract. Al Hallaniyat Island is located in the southern part of Oman. The island has a low number of population and is currently fed by a number of diesel generators with a capacity of 1568 kW. Decreasing the emission of greenhouse gases is one of the main goals for the country's vision of 2040 as well as Oman's commitment to net-zero emissions by 2050, in line with the Paris Agreement of limiting global warming to 1.5°C. The purpose of this paper is to study the feasibility of replacing diesel generators with 100% renewable energy sources. The proposed energy source for the Island will be a hybrid system consisting of the wave energy converter, PV, wind and battery storage system. Real load data was used in the model with measured wind speed and solar radiation. The Hybrid Optimization Model for Electric Renewables software is used in this study. Several systems were proposed but the most economical one is PV-Wind-Wave-Battery system that resulted in levelized cost of energy of US$/kWh 0.1586 and Net Present Cost (NPC) of US$ 5,010,594 . Having PV-Wave-Battery will result in levelized cost of energy of US$/kWh 0.174 whereas having Wind-Wave-Battery will increase the energy cost to US$/kWh 0.42.

1 Introduction

Wave energy is obtained from winds as they blow across the seas. The power in a wave depends on square of the wave amplitude and period. Thus, large amplitude (around 2 m) and long wave (7-10 second) having energy that ranges between 40-70 kW per m width of coming wave [1]. The energy intensity of the wave decreases near the coast due to interaction with the seabed. Marine energies are of untapped resources when compared with solar and wind energies in countries that have seas and oceans. Wave energy has the highest energy density among other marine renewable energy sources [2]. Several studies explored the possibility of utilizing wave energy in various locations around the world, such as Gulf of Oman [3,4], the Persian Gulf [4,5], the Atlantic coast of Europe [6-7], Atlantic Ocean [8] and many other locations. There are several advantages of utilizing

* Corresponding author: albadi@squ.edu.om

wave energy such as, having limited environmental impact, the energy is available throughout the day, and having high-energy fluxes.

Most modern countries worldwide are working hard to shift their power generation away from fossil fuels and toward renewable sources. The environmental consequences associated with producing power with fossil fuels, as well as their storage limitations, are particularly troubling for our human beings [9]. Coal is the world's primary fossil fuel used to generate energy [10]. A technical and economic analysis of a hybrid renewable energy system for powering small plant auxiliary units was presented by many researchers [11]. Electrical system developers and legislators identified integrated units to be highly beneficial in distribution systems, which are dependent on the total characteristics of each unit, such as solar power systems, wind systems, load characteristics, and local renewables [12].

The increase in energy demand in Oman, which mainly depends on fossil fuels, causes more emission of gases as well as a depletion of fossil-fuel reserves [13]. Oman is blessed with a high level of renewable energy resources (Solar, Wind and wave). The solar density in Oman is among the highest in the world in all governorates [14-16]. Wind is another good potential for renewable energy, particularly in Oman's coastal and southern regions [17-20].

Wave energy is another undeveloped renewable energy source. Oman, with over 2000 kilometres of maritime border with the Indian Ocean, Arabian Sea and the Gulf of Oman, possesses a potential and relentless source of wave energy that calls for special attention. Sea waves are among the largest potent energy carriers in renewable energy sources since they have abundant energy resources in all geographical zones. Scientists believe that around the world, energy generated from the ocean waves can reach up to 2 Tearawatts (TW) per year. The total theoretical energy from waves equals to 8*10^6 Terawatt hours (TWh) per year, meaning it's approximately 100 times the total hydroelectric power of the entire plant [21]. Producing this electricity with fossil fuels would result in 2 million tons of CO_2 emissions. The International Energy Agency (IEA) predicts that fossil fuels contribute to around 80% of gross primary energy by 2030, assuming a 1.6% annual increase in world energy consumption [22]. Wave energy, when used in accordance with the Kyoto Protocol, can help to reduce air pollution [21]. The development of wave energy conversion has been ongoing for many years. Models and sizes of the produced technology vary which results in expensive and time-consuming operations for big and complex machinery. As we all know, wave motion is highly arbitrary. As a result, a mechanism is required to use and convert the erratic input to a consistent and steady output in the form of electricity. Wave forces are absorbed by an absorber. This absorbed energy can be transformed immediately into electricity using a Linear Permanent Magnet Generator or in stages utilizing Hydraulic and Mechanical transducers. Some factors to consider when selecting a converter include durability, longevity, efficiency and reliability [23].

Al Hallaniyat Island utilizes diesel generators for energy generation. Due to massive investments requirement, extending the grid to the Island will be uneconomical and technically very difficult [24]. The aim of this paper is to study the possibility of utilizing wave, solar and wind energy with battery storage system to replace diesel power generation. Due to their balancing nature and low carbon emissions, the combination of wave, wind and solar energy provides reliable and cost-effective options for forming a hybrid renewable energy system for power generation in the Island.

2 Existing System

2.1 Generations

Diesel generators are currently utilized to provide the energy requirement for the Island; the total capacity is around 1568 kW.

2.2 Load Profile

The monthly peak load profile for Al Hallaniyat Island for the year 2021 is presented in Figure 1 [25]. The demand during summer time is low owing to the monsoon season, the demand increases during the month of May to 485 kW and 445 kW in September.

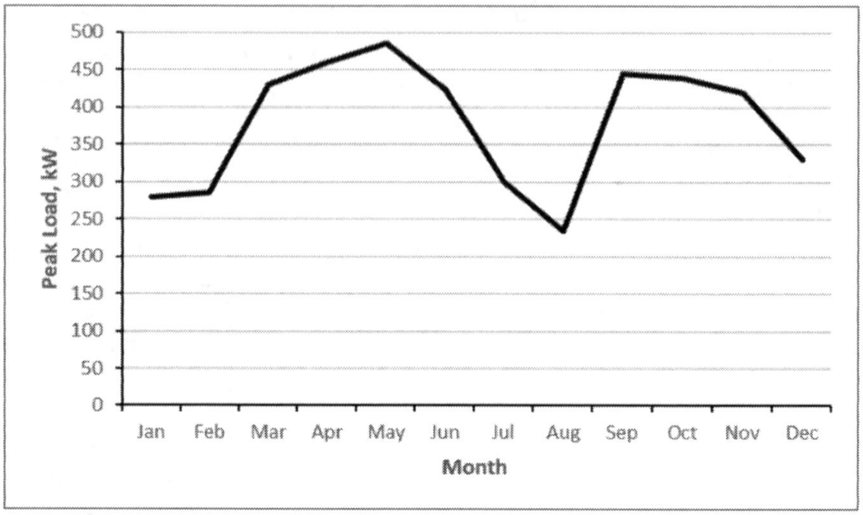

Fig. 1. Measured Monthly Peak Load at Al Hallaniyat Island, 2021

2.3 Solar Irradiance

The measured result for the solar irradiancefor the year 2022 is shown in Figure 2. The Directorate General of Meteorology, Civil Aviation Authority [26], provides these data for a site close to Al Hallaniyat Island.

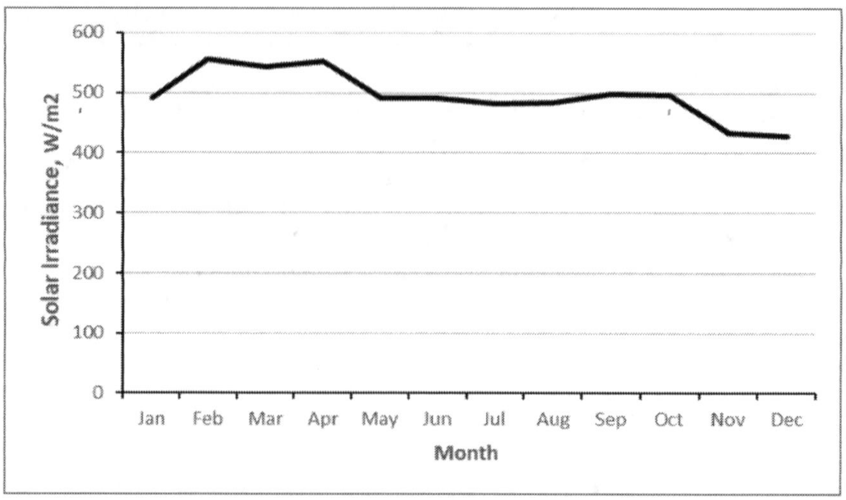

Fig. 2. Measured Monthly Average Daily Global Radiation at a Site (Shalim) Near Al Hallaniyat Island, 2022

2.4 Wind Speed

The wind speed during the monsoon season is high reaching more than 10 m/s, however the wind speed drops to around 3 m/s during wintertime as depicted in Figure 3. The wind speed was measured at 10 m anemometer height and provided by the Directorate General of Meteorology, Civil Aviation Authority [26].

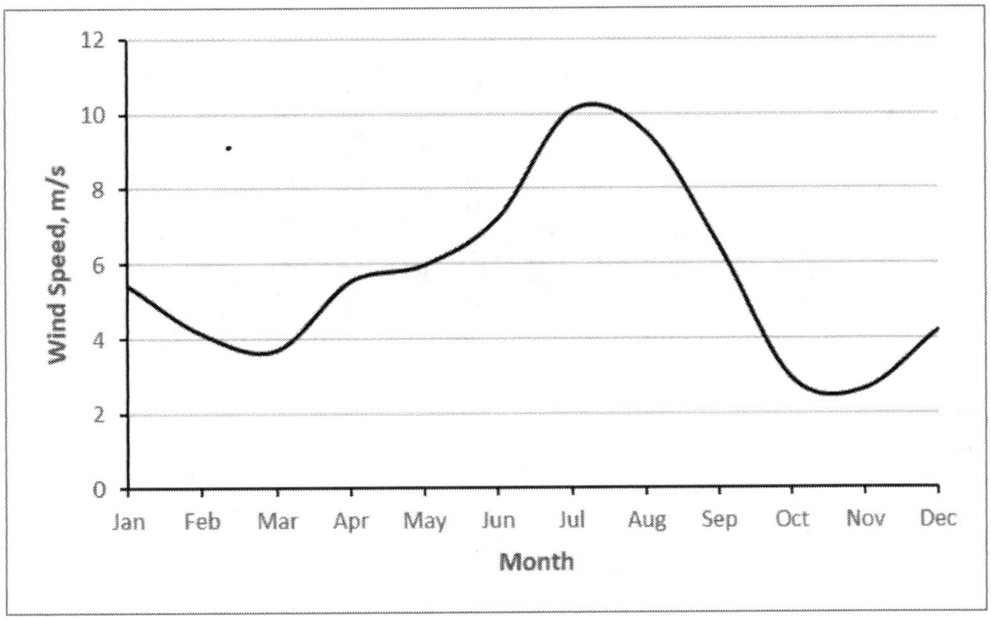

Fig. 3. Measured Monthly Average Daily Wind Speed at Al Hallaniyat Island, 2022

2.5 Wave Energy

Figure 4 represents the average monthly wave heights near Al Hallaniyat Island obtained from the Directorate General of Meteorology, Civil Aviation Authority [26]. During the monsoon season, high wind speed creates the highest wave height with high-speed wave, as indicated in Figure 5. The wave speed is calculated using equation 1 [27]:

$$\text{Wave Speed}=1.56*T \quad (1)$$

Where:

T is the wave period in seconds

https://doi.org/10.1051/e3sconf/202451403001

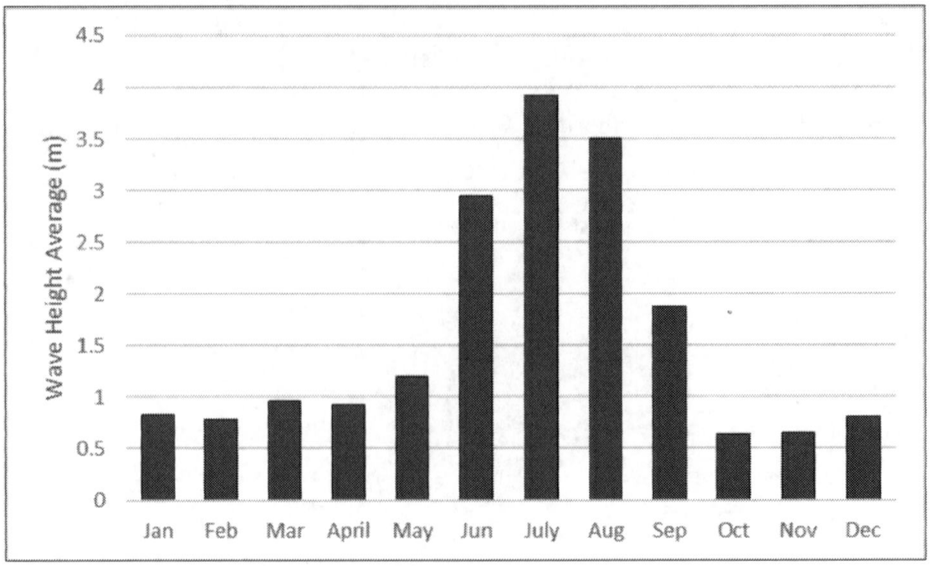

Fig. 4. The Monthly Average Wave Height Data

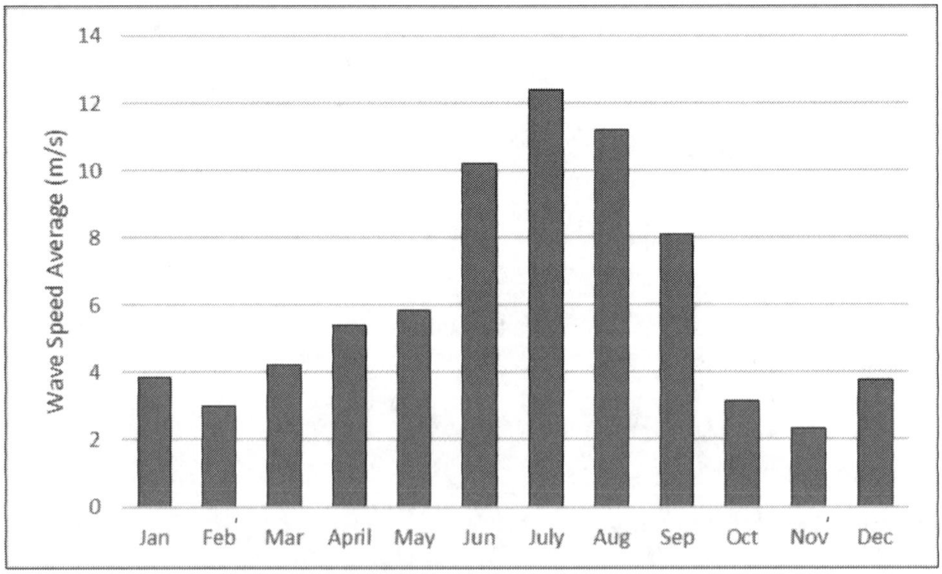

Fig. 5. The Monthly Average Wave Speed Data

The power of the wave can be calculated using the following equation 2 [28]:

$$P=H^2T \quad (2)$$

Where, H is the wave height in meter, T is the wave period in second and P is the wave power in kW/m. Figure 6 presents the monthly average power (kW/m). The peak power occurs during the period from June to September owing to the high wind speed.

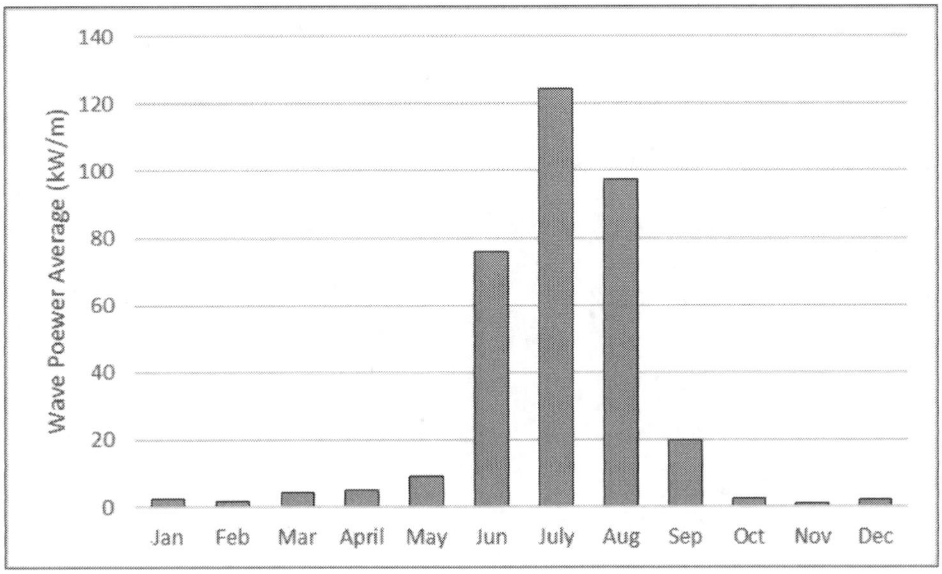

Fig. 6. The Monthly Average Power Output (kW/m) Trend Line

3 Modeling of the Proposed System Using Homer

Replacing diesel engines with renewable energy sources will reduce greenhouse gas emission. To meet the load demand from only renewable it is recommended to install a hybrid system consisting of PV, wind and wave energy converter since all of them are promising resources, and with energy storage. The schematic of the proposed hybrid system is shown in Figure 7. Solar PV generates electricity only during the day whereas wind turbines can produce energy when there is sufficient wind speed during day and night. Wave energy converter generates energy during the wave movement, which happens during the whole day. The cost and technical data for various components of the hybrid system are presented in Table 1. The whole system was modeled using HOMER Pro software [29].

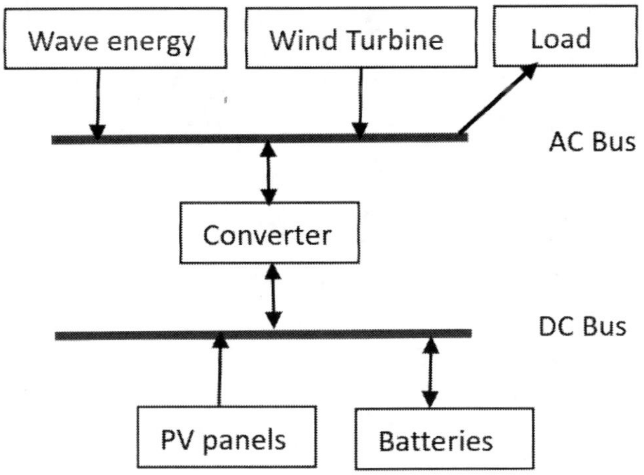

Fig. 7. Schematic Diagram of the Proposed System (PV-Wind-Wave Converter-Batteries)

Description	Data	Description	Data
PV		*Wave converter [36]*	
Capital cost	1000 US $/kW	Power rating	70 kW
Lifetime	25 years	Life time	25 years
O&M cost	10 US $/kW/year	Capital cost	6000-2200$/kW
Wind		Replacement cost	5750-2000 $/unit
Capital cost	900 US $/kW	Maintenance cost	3% of initial cost/year
Lifetime	25 years		
O&M cost	36 US $/kW/year		
Inverter			
cost	250 US $/kW		
Lifetime	15 years		
O&M	0 US $/year		
Diesel unit			
Each unit	250 US $/kW		
Diesel	0.67 US $/Liter		
Discount rate	7.55%		
Inflation rate	2%		
capacity shortage	2%		

Table 1. Cost and Technical Data for PV, Diesel Engine, Wind turbine and Inverter [30-35].

4 Discussion of Results

4.1 Existing System

Al Hallaniyat Island is fed by group of diesel generators, the schematic diagram for the existing system is presented in Figure 8. The energy production in MWh for the whole year is shown in Figure 9. The peak demand occurs in May and October, which are the hot months in the Island. The summertime represents the lowest energy demand due to moist monsoon wind blowing from the Indian Ocean towards the land.

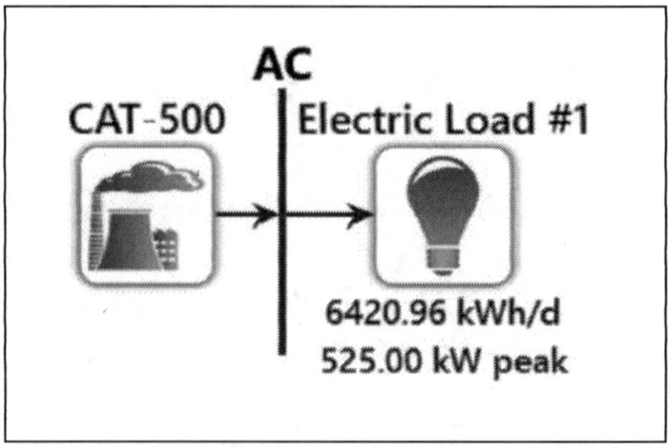

Fig. 8. Schematic Diagram for the Existing System [29]

https://doi.org/10.1051/e3sconf/202451403001

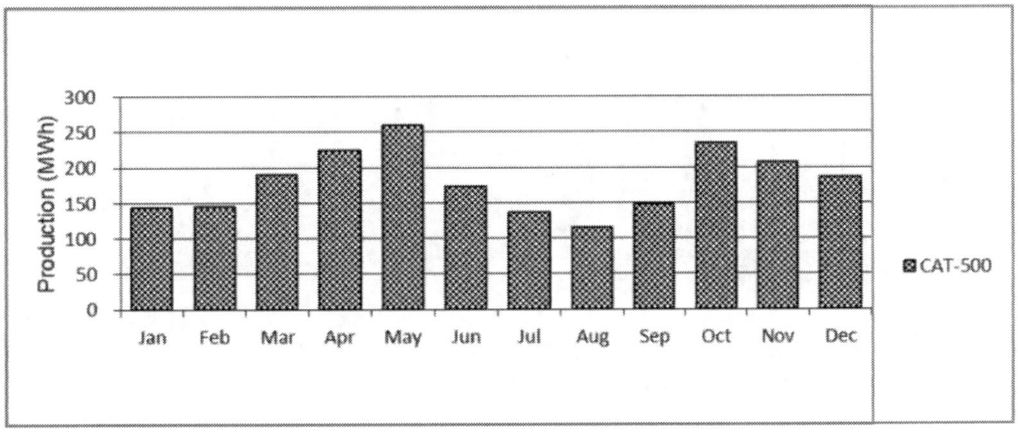

Fig. 9. The Monthly Energy Production of Al Hallaniyat Power Station

4.2 The Hybrid System

The schematic diagram for the proposed system is shown in Figure 10, which consists of a Wave energy converter, PV, Wind and storage batteries. In this system, the generation sources which are 100% renewable serve the electrical load. The Monthly Energy Production for the system is presented in Figure 11. In the proposed system, the wave energy converter, PV and Wind sources serve the electrical load with large contribution of PV. Both PV and Wind will provide energy during the whole year whereas the wave energy converter provides energy for eight months and can provide the energy for the whole day. The wave energy converter [70kW] has a total annual production of 323,792 kWh/yr. It contributes with 13.8% of energy production, has a capacity factor of 52.8% and has a levelized cost of energy of 0.209 $/kWh as shown in Table 2. Noteworthy, removing wind or PV sources from the system will increase the levelized cost of energy.

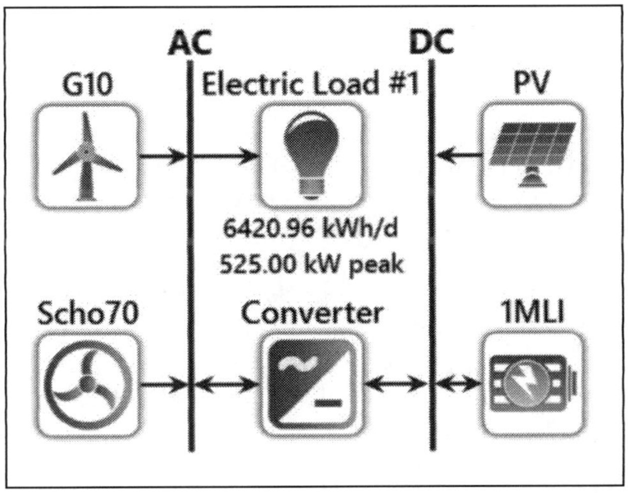

Fig.10. Schematic Diagram for Wave energy converter with PV, Wind and Batteries [29]

https://doi.org/10.1051/e3sconf/202451403001

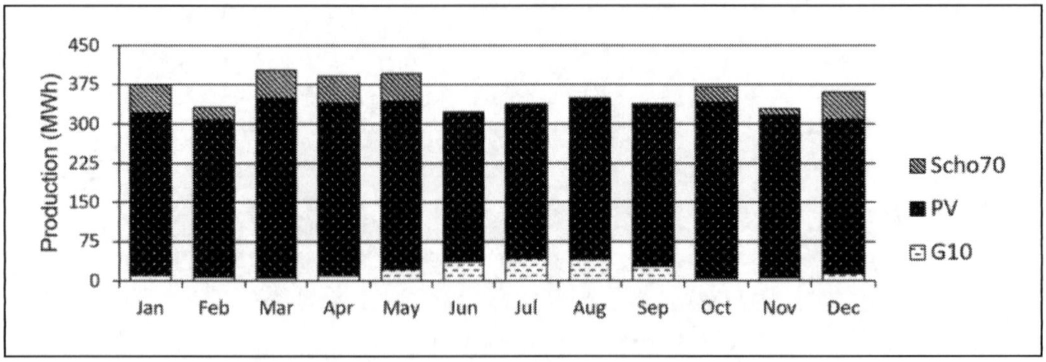

Fig. 11. The Monthly Energy Production for PV-Wind and Batteries Station

Table 2. The Wave Energy Converter Parameters

Wave Energy Converter	Value
Rated Capacity	70.0 kW
Capital Cost	$420,000
Hours of Operation	5,832 hr/yr
Levelized Cost	0.209 $/kWh
Capacity Factor	52.8 %
Total Production	323,792 kWh/yr
Maintenance Cost	12,600 $/yr
Maximum Output	70.0 kW
Hydrokinetic Penetration	13.8%

5 Conclusions

Al Hallaniyat Island is currently fed by diesel generators only, which have high operating costs as well as emission of greenhouse gases. Oman placed a target to increase renewable energy in its energy mix. The target based on Oman's vision should reach 20% by 2023. Several renewable energy large and small projects are already operating. In this paper, we propose to utilize a hybrid energy system (100% renewable) for this Island to meet the load requirement. A combination of PV, Wind, Wave energy converter and batteries are proposed. The Net Present Cost (NPC) is found to be US$ 5,010,594 and the levelized cost of energy is US$/kWh 0.1586. The proposed system cost is high owing to the capital cost of the Wave energy converter. It is expected that the cost of wave system to decline since several countries are planning to utilize this kind of energy source.

The authors would like to acknowledge The Tanweer Company, Oman, who provided reliable support in the form of load and diesel generator data. Additionally, the authors thank the Civil Aviation Authority's Directorate General of Meteorology for its assistance in providing the wind, wave and solar data for Al Hallaniyat Island.

References

1. Alain Cle´ment, et al. "*Wave energy in Europe: current status and perspectives*", Renewable and Sustainable Energy Reviews 6 (2002) 405–431
2. M. Leijon, H. Bernhoff, M. Berg, O. Agren, "*Economical considerations of renewable electric energy production-especially development of wave energy"*, Renew. Energy 8 (2003) 1201-1209.
3. Bahareh Kamranzad, Vahid Chegini, Amir Etemad-Shahidi, *"Temporal-spatial variation of wave energy and nearshore hotspots in the Gulf of Oman based on locally generated wind waves"*, Renewable Energy 94 (2016) 341-352.
4. Mohammad Hossein Jahangir, Ardavan Shahsavari , Mohammad Amin Vaziri Rad , *"Feasibility study of a zero emission PV/Wind turbine/Wave energy converter hybrid system for stand-alone power supply: A case study",* Journal of Cleaner Production 262 (2020) 121250.
5. Mohammad Hossein Jahangir, Samaneh Fakouriyan, Mohammad Amin Vaziri Rad, Hassan Dehghan,*"Feasibility study of on/off grid large-scale PV/WT/WEC hybrid energy system in coastal cities: A case-based research*", Renewable Energy 162 (2020) 2075-2095
6. M. Gonçalves, P. Martinho, C. Guedes Soares, *Wave energy conditions in the western French coast,* Renew. Energy 62 (2014) 155-163.
7. J.P. Sierra, D. G. Marco, J. Sospedra, X. Gironella, C. M., A. S, Arcilla, "*Wave energy resource assessment in Lanzarote (Spain)*", Renew. Energy 55 (2013) 480-489.
8. Bryson Robertsona, Jessica Bekkerb, Bradley Buckhamb, "*Renewable integration for remote communities: Comparative allowable cost analyses for hydro, solar and wave energy*", Applied Energy 264 (2020) 114677.
9. T. Ali, H. Ma, and A. J. Nahian, "*Techno-Economic Analysis of a Hybrid Mini-grid in Rural Areas: A Case Study of Bangladesh,*" J. Energy Res. Rev., vol. 4, no. 1, pp. 10–29, 2020.
10. Barua, Prosenjit & Ghosh, Bikram. (2021). *Feasibility Analysis of Renewable Energy Based Hybrid Power System in a Coastal Area, Bangladesh.* 10.1109/ETCCE51779.2020.9350875.
11. S.P. Makhija and S. P. Dubey, (2016), *Optimally Sized Hybrid Energy System for Auxiliaries of A Cement Manufacturing Unit with Diesel Fuel Price Sensitivity Analysis*, Taylor & Francis, International Journal of Ambient Energy, 1-6.
12. Adefaratiand, R. C. Bansal (2016), "*Integration of Renewable Distributed Generators into the Distribution System: A Review*", IET-Renewable Power Generation, Vol. **10**, No. 7, pp. 873-884.
13. M. Abujubbeh, V. T. Marazanye, Z. Qadir, M. Fahrioglu and C. Batunlu, "*Techno-Economic Feasibility Analysis of Grid-Tied PV-Wind Hybrid System to Meet a Typical Household Demand: Case Study - Amman, Jordan*," 2019 1st Global Power, Energy and Communication Conference (GPECOM), Nevsehir, Turkey, (2019), pp. 418-423.
14. A. H. Al-Badi "*Performance Assessment of 20.4 kW Eco-House Grid connected PV Plant in Oman*", Journal of Sustainable Engineering, Volume **13**, (2020) - Issue 3, pp. 230-241.
15. A. H. Al-Badi, "*Measured Performance Evaluation of a 1.4 kW Grid connected desert type PV in Oman*" Energy for Sustainable Development, Volume **47**, December (2018), Pages 107-113.
16. H. Yousef, A. H. Al-Badi, A. Polycarpou "*Power Management for Hybrid Distributed Generation Systems*", International Journal of Sustainable Energy, 11:1, 2018, 65-74.
17. A. Al-Wahabi and A. Al-Badi, *" Sizing of a Hybrid Energy System for Al Mazyouna Area "*, Renewable Energy and Power Quality Journal, Volume **20**, September (2022), Pages 438-443. 10.24084/repqj20.335

18. Razzaqul Ahshan, Ahmet Onen, Abdullah H. AlBadi, "*Assessment of Wind-to-Hydrogen (Wind-H2) Generation Prospects in the Sultanate of Oman*" Renewable Energy Journal, Volume **200**, November (2022), Pages 271-282
19. Abdullah Al-Badi, Abdulmajeed Al Wahaibi, Razzaqul Ahshan and Arif Malik,"*Techno-Economic Feasibility of a Solar-Wind-Fuel Cell Energy System in Duqm, Oman*", Energies (2022), 15, 5379.
20. IRENA (2014) Renewables readiness assessment sultanate of Oman, Renewables Readiness Assessment Sultanate of Oman. Available at: https://prodcd.irena.org/Publications/2014/Nov/Renewables-Readiness-Assessment-Sultanate-of- Oman (Accessed: August 28, 2023).
21. L. Rodrigues, Wave power conversion systems for electrical energy production, RE&PQJ 1 (6) (2008).
22. S.H. Salter, Wave power, Nature 249 (5459) (1974) 720–724.
23. Veerabhadrappa, Kavadiki & B.G, Suhas & Mangrulkar, Chidanand & Kumar, R. & Mudakappanavar, V.S. & Kambalur, Narahari & Seetharamu, Nagashree. (2022). *Power Generation Using Ocean Waves: A Review*. Global Transitions Proceedings. 3. 10.1016/j.gltp.2022.05.001.
24. A. Haghighat Mamaghani, S. A. Avella Escandon, B. Najafi, A. Shirazi, and F. Rinaldi, "*Techno-economic feasibility of photovoltaic, wind, diesel and hybrid electrification systems for off-grid rural electrification in Colombia*," Renew. Energy, vol. **97**, pp. 293–305, (2016).
25. Tanweer Report, (2021). https://tanweer.om/uploadsall/Pdfs/9d7823e8-6cf8-4aaa-bb59- b88f843fad2e.pdf
26. Directorate General of Meteorology, Civil Aviation Authority. https://www.caa.gov.om/caa/directorates/directorate-general-of-meteorology.
27. Graham Warren," *Basic Wave Theory Review*", Bureau of Meteorology Australia, power, (2003).
28. Boehme, T., 2006. Matching renewable electricity generation with demand in Scotland (Doctoral dissertation, University of Edinburgh). https://webarchive.nrscotland.gov.uk/3000/https://www.gov.scot/Publications/2006/04/24110728/10.
29. HOMER Pro Version 3.14.2. Available online: https://www.homerenergy.com/products/pro/index.html
30. Shahid Hussain Siyal, Dimitris Mentis, and Mark Howells. *Economic analysis of standalone wind-powered hydrogen refueling stations for road transport at selected sites in Sweden*. International Journal of hydrogen energy 40, (2015): 9 8 5 5-98 6 5.
31. Vikas Khare, Savita Nema, and Prashant Baredar. *Optimization of hydrogen based hybrid renewable energy system using HOMER, BB-BC and GAMBIT.* International Journal of hydrogen energy **41**, (2016): 16743-16751.
32. Nader Barsoum, Pearl Dianne Petrus. *Cost Optimization of Hybrid Solar, Micro-Hydro and Hydrogen Fuel Cell Using Homer Software*. Energy and Power Engineering, (2015), **7**, 337-347
33. Khare, V., S. Nema, and P. Baredar, *Optimization of hydrogen based hybrid renewable energy system using HOMER, BB-BC and GAMBIT.* International Journal of Hydrogen Energy, (2016). 41(38): p. 16743-16751.
34. Barsoum, N. and P.D. Petrus, *Cost optimization of hybrid solar, micro-hydro and hydrogen fuel cell using homer software*. Energy and Power Engineering, (2015). 7(08): p. 337.
35. Bosserelle, C., Reddy, S.K., Krüger, J. *Cost Analysis of Wave Energy in the Pacific*. Pacific Community. 2016

36. Daniel Friedrich, George Lavidas, "*Evaluation of the effect of flexible demand and wave energy converters on the design of hybrid energy systems*", IET Renew. Power Gener., 2017, Vol. 11 Iss. 9, pp. 1113-1119

 https://doi.org/10.1051/e3sconf/202451403002

On the Efficiency of Solar Radiation Conversion into Steam, Production of Clean Water, and Generation of Electrical Energy

Quoc Thinh Tran [1*], *Alexander* Dmitriev[1], *Inna* Mikhailova[1]

[1]National Research University "Moscow Power Engineering Institute", 111250 Moscow, Russia

Abstract. This study focuses on the development of a model for evaluating the efficiency of converting solar radiation into thermal energy. The efficiency coefficient is represented as a function of multiple variables, including system geometry, time, properties of the working fluid, and external parameters. The developed model is a simplified version and is applicable to systems with a stationary working fluid undergoing heating and vaporization due to solar radiation. Calculations were performed based on the model using a gold-water nanofluid. The results of the calculations demonstrated the existence of an optimal range of optical thickness and nanoparticle concentration, where the working fluid is effectively heated and vaporized. It is shown that the efficiency of systems employing nanofluids as heat transfer fluids cannot be evaluated solely based on the absolute value of the heating efficiency coefficient. However, it can be assessed through its derivative with respect to time. The faster the heating efficiency coefficient decreases over time, the more efficient the system is in terms of heating, and the sooner it reaches a steady-state condition. The developed model serves as a foundation for the advancement of more sophisticated models that allow for the evaluation of various other factors, such as complex geometries with forced fluid flow, the nature of interactions between nanoparticles and the base fluid, as well as mechanisms of solar radiation conversion into thermal energy.

1 Introduction

Currently, significant attention is being devoted to research in the field of green energy, including the application of nanomaterials to enhance the efficiency of energy systems [1-2]. Due to their improved thermal and optical properties, nanomaterials are being used in energy applications, particularly in solar thermal energy, as a new class of heat transfer fluids for steam generation, clean water production, and electricity generation [3-5]. However, in recent studies, the efficiency of converting solar radiation into steam has primarily been evaluated based on the thermal balance equation, where it depends solely on the heat of vaporization and the incident solar radiation power [1, 6-7]. Yet, the results of numerous experimental investigations have shown that the efficiency of converting solar radiation into steam or

* Corresponding author: tranqth.96@gmail.com

electrical energy significantly depends on various factors [8-9]. Moreover, experimental data on the efficiency of energy systems utilizing working fluids such as nanofluids exhibit such a wide scatter that it becomes challenging to evaluate and apply the obtained results for practical purposes as a whole. Therefore, as part of the scientific project No. 23-19-00840, supported by the Russian Science Foundation, we propose to systematize and develop a mathematical model for calculating the efficiency of solar radiation conversion using nanomaterials for steam generation, clean water production, and electricity generation through various methods, including thermoelectric conversion and other types of conversion. In the initial stage, a simplified model has been developed, which allows for the assessment of the appropriateness of using different types and concentrations of nanomaterials, irradiation geometry, solar radiation intensity, and properties of the base fluids in various energy and environmental technologies.

2 Mathematical model for the heating and vaporization efficiency of nanofluids

2.1 Model of transient temperature field

Typically, when solar radiation is absorbed, the working fluid initially heats up, and its temperature increases over time. The effective evaporation process occurs only after the entire system reaches a steady-state condition. To assess the efficiency of heating and evaporation processes of nanofluids under the influence of solar radiation, this study considers a one-dimensional problem, which is schematically represented in Fig. 1.

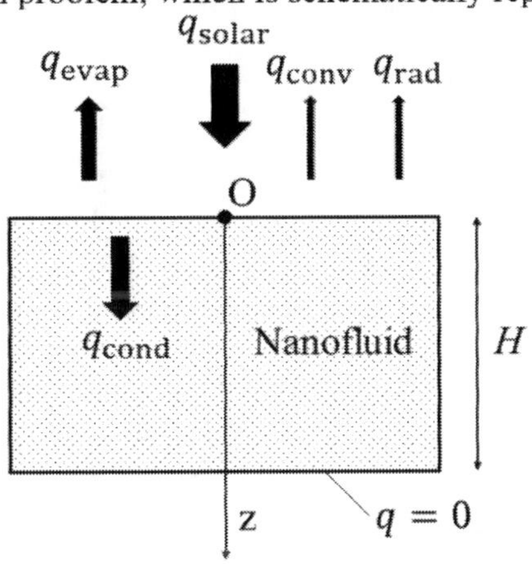

Fig. 1. Physical model for assessing the efficiency of converting solar radiation into thermal energy in a one-dimensional problem formulation.

At the interface between the nanofluid and air phases, incident solar radiation (q_{solar}) is absorbed and converted into thermal energy. Initially, during the transient process, a significant portion of the solar radiation energy is converted into thermal energy and transferred through conduction (q_{cond}) and convection to the deeper layers of water. In the steady-state regime, a major portion of the generated thermal energy is carried away from the interfacial surface through latent heat of vaporization (q_{evap}) during the evaporation process. The remaining energy is lost to the surrounding environment through air convection (q_{conv}) and radiation from the interfacial surface (q_{rad}). The heat exchange in nanofluids under the influence of solar radiation is a complex process, which is described by the following differential equation of energy conservation

$$\rho_{nf} \cdot c_p \cdot \left(\frac{dT}{dt} + \mathbf{w} \cdot \text{grad}T\right) = \lambda_{eff} \nabla^2 T - \text{div}(\mathbf{q}_R) + q_V, \quad (1)$$

where ρ_{nf}, c_p, λ_{eff} are the density, specific heat capacity, and effective thermal conductivity of the working nanofluid, respectively; **w** is the velocity vector of the medium; q_V is the internal heat source; $\mathbf{q}_R$ is the heat flux density vector due to thermal radiation; *T* is the temperature; and *t* is the time.

In the case of a low-intensity heating and evaporation process, in a first approximation, free convection within the volume of the nanofluid can be neglected, i.e., w = 0. To describe the process of radiation energy transfer in one-dimensional space, the following integral relationship can be assumed

$$q_R(z) = q_{solar} \cdot \exp(-\sigma_{nf} \cdot z), \tag{2}$$

where $q_R(z)$ is the component of the heat flux density vector of thermal radiation within the volume of the nanofluids in the z-direction; q_{solar} represents the heat flux density of thermal radiation incident on the surface of the nanofluid, W/m²; σ_{nf} denotes the integrated effective extinction coefficient of the nanofluid, m⁻¹.

In the formulation of the one-dimensional problem without an internal heat source ($q_V = 0$) and considering the adopted assumptions, equation (1) and its corresponding uniqueness conditions can be expressed as follows

$$\rho_{nf} \cdot c_p \cdot \frac{dT}{dt} = \lambda_{eff} \cdot \frac{\partial^2 T}{\partial z^2} + \sigma_{nf} \cdot q_{solar} \cdot \exp(-\sigma_{nf} \cdot z), \tag{3}$$

$$T(z,t)|_{t=0} = T_0, \tag{4}$$

$$-\lambda \cdot \left.\frac{dT}{dz}\right|_{z=0} = q_{\text{conv}} + q_{\text{evap}} + q_{\text{rad}}, \tag{5}$$

$$-\lambda \cdot \left.\frac{dT}{dz}\right|_{z=h} = 0, \tag{6}$$

where T_0 is the initial temperature of the system in degrees Celsius. The thermophysical properties (ρ_{nf}, c_p, λ_{eff}) of nanofluids are determined using the relationships presented in Ref. [10], while their optical properties (σ_{nf}) of the nanofluids are determined using the relationships of Rayleigh scattering theory [11]. Expression (6) defines the heat insulation condition at the bottom of the working fluid layer. On the other hand, equation (5) represents a boundary condition of the second type, where the heat flux density is specified at the interfacial surface between the nanofluid and the air. In this case, the heat flux densities on the right-hand side of equation (5) are determined according to the following relationships

$$q_{\text{evap}} = j \cdot L, \tag{7}$$

$$q_{\text{conv}} = \alpha_{\text{conv}} \cdot (T_s - T_{amb}), \tag{8}$$

$$q_{\text{rad}} = \varepsilon\sigma(T_s{}^4 - T_{amb}{}^4), \tag{9}$$

here, *j* represents the mass flow rate density during evaporation, kg/(m²·s); *L* is the latent heat of vaporization, J/kg; α_{conv} is the heat transfer coefficient between the interfacial surface and the surrounding medium, W/(m²·K); ε and T_s are the emissivity and temperature of the interfacial surface, respectively; T_{amb} is the temperature of the ambient medium far from the interfacial surface, °C.

Since during the non-steady heating period of the nanofluid, a significant portion of the radiation energy is used to increase its temperature, it is proposed to estimate the mass flow rate density *j* due to evaporation solely based on diffusion through the following relationship

$$j = -M \cdot D \cdot \nabla c = M \cdot D \cdot \frac{C_{sat} - C_\infty}{\Delta z}, \tag{10}$$

where *M* is the molar mass of water vapor, which is equal to $18 \cdot 10^{-3}$ kg/mol; *D* is the diffusion coefficient of water vapor in air, which is equal to $2.1 \cdot 10^{-5}$ m²/s under normal conditions; Δz is the distance from the interfacial surface where the vapor concentration is equal to C_∞. To calculate *j*, it is proposed that a constant relative humidity of 40% is maintained at a distance of 1 m from the interfacial surface, which corresponds to $C_\infty = 0.222$ mol/m³. C_{sat} is the saturation concentration of water vapor directly at the interfacial surface, which is determined using the following formula

$$C_{sat} = \frac{P_{sat}}{R \cdot [273.15 + T_s] \cdot M'} \tag{11}$$

$$P_{sat} = 133 \cdot 10^{8.07131 - \frac{1730.63}{233.426 + T_s}}, \tag{12}$$

here, R is the individual gas constant for water vapor, which is equal to 461.9 J/(kg·K); P_{sat} is the saturation pressure of water, which corresponds to the temperature of the interfacial surface T_s.

2.2 Model of the stationary temperature field

Neglecting heat losses due to radiation and assuming the relationship for the heat flux density due to evaporation as $q_{evap} = \alpha_{evap} \cdot (T_s - T_{amb})$, the stationary temperature field can be obtained analytically by solving the system of equations (3)-(6) and has the following form

$$T(z) = -\frac{q_{solar}}{\lambda_{eff} \cdot \sigma_{nf}} \cdot \exp(-\sigma_{nf} \cdot z) + C_1 \cdot z + C_2, \tag{13}$$

$$C_1 = -\frac{q_{solar}}{\lambda_{eff}} \cdot \exp(-\sigma_{nf} \cdot H), \tag{14}$$

$$C_2 = T_{amb} + \frac{q_{solar}}{\lambda_{eff} \cdot \sigma_{nf}} + \frac{\lambda_{eff}}{\alpha_{conv} + \alpha_{evap}} \left(\frac{q_{solar}}{\lambda_{eff}} + C_1 \right), \tag{15}$$

here, C_1 and C_2 are integration constants that depend on the properties of the nanofluids (λ_{eff}, σ_{nf}), the intensity of incident radiation (q_{solar}), and the geometry of the system (H); α_{evap} is the heat transfer coefficient of the evaporation process, which is conventionally assumed to be 40 W/(m^2·K) when interacting with natural solar radiation [6].

2.3 Efficiency of solar radiation conversion into thermal energy

During the non-steady period, the efficiency of the solar radiation conversion process into useful thermal energy (η_t), including the heating (η_{heat}) and vaporization (η_{evap}) of nanofluids, is determined by the following relationships

$$\eta_T = \eta_{heat} + \eta_{evap}, \tag{16}$$

$$\eta_{heat} = \frac{c_p(T_{mid} - T_0) \cdot (m_0 - \Delta m)}{F \cdot q_{solar} \cdot \Delta t}, \tag{17}$$

$$\eta_{evap} = \frac{L \cdot \Delta m}{F \cdot q_{solar} \cdot \Delta t'} \tag{18}$$

where T_0 and m_0 are the initial temperature and mass of the liquid volume, respectively; Δm is the amount of evaporated liquid during the time interval Δt; F is the area of the interfacial surface between the liquid and air; T_{mid} is the average temperature of the liquid volume after the time of exposure Δt and is determined by the formula

$$T_{mid} = \frac{1}{H} \int_0^H T(z,t) \mathrm{d}z,$$

$$\Delta m = j \cdot F \cdot \Delta t,$$

$$m_0 = \rho \cdot F \cdot H.$$

In the case of $\Delta m \ll m_0$, we obtain the following

$$\eta_t = \frac{c_p(T_{mid} - T_0) \cdot \rho \cdot H + j \cdot L \cdot \Delta t}{q_{solar} \cdot \Delta t}. \tag{19}$$

From this expression, it follows that the efficiency of converting solar radiation into useful thermal energy depends on parameters such as the liquid layer thickness (H), the average temperature of the liquid volume (T_{mid}), and the mass flow rate density during evaporation (j). The last two quantities, in turn, depend on the thermophysical and optical properties of the nanofluids, namely, the characteristic size and shape of the nanoparticles, their concentration and material type, as well as the environmental conditions (T_{amb}, C_∞). Therefore, η_t can be represented as a function of multiple variables, i.e.

$$\eta_t = f(H, t, q_{solar}, \phi, \psi, d, T_s, T_{amb}, C_\infty). \quad (20)$$

where ϕ is the volume concentration of nanoparticles; ψ is the sphericity, which is equal to unity for spherical nanoparticles; d is the diameter of the nanoparticles.

3 Results and discussion

3.1 The efficiency of heating nanofluids

The developed model was constructed using Mathcad software, based on which the dependence of η_t on parameters (H, t, ϕ) was investigated for a gold-water nanofluid in the non-steady process of solar radiation conversion into thermal energy. The following conditions were considered: $\psi = 1$, $d = 20$ nm, $q_{\text{solar}} = 1000$ W/m^2, $T_{\text{amb}} = T_0 = 20$°C, $C_\infty = 0.222$ mol/m^3. The variation of the temperature field with respect to parameters (H, t, ϕ) is shown in Fig. 2, and the dependence of $\eta_t(H, t, \phi)$ is presented in Fig. 3.

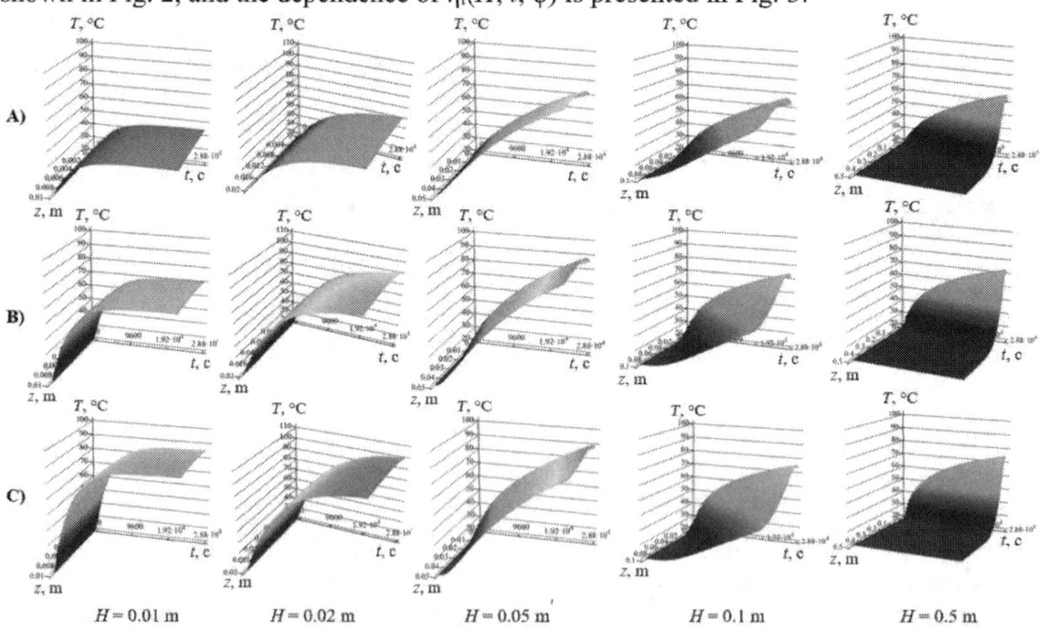

Fig. 2. The variation of the temperature field $T(z)$ of gold-water nanofluid over time is shown, depending on the volume concentration of Au nanoparticles (A. $\phi = 0$ ppm, B. $\phi = 50$ ppm, C. $\phi = 100$ ppm) and the height of the liquid layer (H) at $q_{\text{solar}} = 1000$ W/m^2.

It can be observed that there exist optimal values of ϕ and H, at which the system can reach a steady-state regime and heat up to a high temperature. Among the examined temperature field calculation options, the optimal parameters ϕ and H are found to be 100 ppm and 0.02 m, respectively. Additionally, from Fig. 2, the following temperature field dependencies on the parameters ϕ and H can be observed. Firstly, as the volume concentration of nanoparticles increases, the maximum temperature $T(z)$ also increases. It can be noticed that when H is in the range of 0.05 to 0.5 m, the temperature field does not differ significantly with an increase in the gold nanoparticle concentration from 50 to 100 ppm (Fig. 2B, C). This is because the gold-water nanofluid with such concentrations nearly

 https://doi.org/10.1051/e3sconf/202451403002

fully absorbs solar radiation at H equal to 0.05 m (Fig. 3C). Secondly, under the influence of natural solar radiation ($q_{solar} = 1000\ W/m^2$), the volume of the gold-water nanofluid can reach a steady-state regime only at small thicknesses of H (0.01 m and 0.02 m). Furthermore, for each nanoparticle concentration, there exists an optimal H thickness at which the working fluid volume reaches the highest temperature. For example, for ϕ equal to 0 ppm, the optimal value of H is 0.05 m, while for ϕ equal to 50 ppm and 100 ppm, the optimal value of H is 0.02 m.

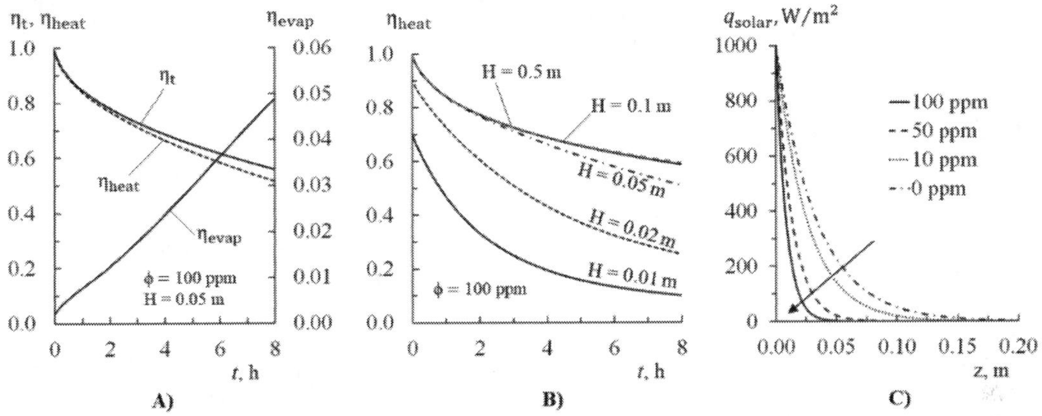

Fig. 3. The variation of heating and evaporation efficiency over time at $\phi = 100$ ppm and $H = 0.05$ m (A), for different values of H (B). Attenuation of radiation intensity across the liquid layer thickness for different concentrations of Au nanoparticles (C).

The calculation results of the efficiency coefficient indicate that during the non-steady heating regime, almost all of the radiant energy is converted into thermal energy to raise the temperature of the working fluid volume (Fig. 3A). As the volume and interfacial surface temperature increase, the heating efficiency gradually decreases. Firstly, this is due to the fact that the heating rate slows down over time, and the system gradually approaches a steady-state regime. Secondly, the rise in interfacial surface temperature leads to more intense evaporation and increases heat losses due to air convection and surface radiation. It is important to note that a high efficiency coefficient does not necessarily imply high hot water or steam production. This can be observed from the analysis of the graph in Fig. 3B and the temperature field images in Fig. 2. For thick layers of the working fluid, such as 0.1 m and 0.5 m, the initial value of the heating efficiency coefficient approaches unity, which is related to its absorbance (Fig. 3C). Over time, the heating efficiency coefficient for systems with thick fluid layers also decreases at a slower rate compared to other cases with smaller H values. This factor indicates that systems with larger H values take longer to heat up and struggle to reach a steady-state regime, which in turn leads to inefficient evaporation. Therefore, the faster the heating efficiency coefficient decreases, the sooner the system reaches a steady-state regime and the more efficiently it generates steam.

Fig. 4A presents the calculation results of the influence of ambient temperature on the heating efficiency coefficient under constant other conditions. It is found that when T_{amb} is higher than T_0, the efficiency coefficient η_{heat} increases, whereas when T_{amb} is lower than T_0, η_{heat} decreases. In the case of T_{amb} being higher than T_0, there exists a region in the initial period of the non-steady heating process where η_{heat} is greater than unity. This is due to the fact that under such conditions, an additional heat flux is supplied to the working fluid volume from the surroundings. In addition to the influence of thickness H and nanoparticle concentration ϕ, the heating rate of the working fluid volume also significantly depends on the intensity of the incident radiation flux q_{solar}. The calculations have demonstrated that with an increase in q_{solar} from 1 to 5 kW/m^2, the heating time of the gold-water nanofluid volume

 https://doi.org/10.1051/e3sconf/202451403002

is reduced by approximately 18 times, as illustrated in Fig. 4B. Further increase in q_{solar} leads to a linear growth of the heating rate.

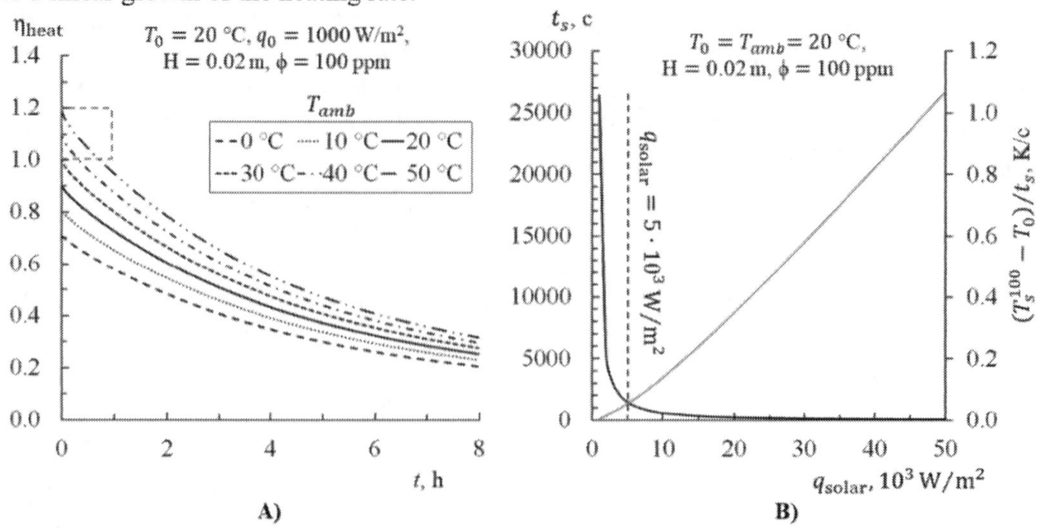

Fig. 4. The influence of ambient temperature on the heating efficiency coefficient (A). The influence of the intensity of the incident radiation flux on the average heating rate of the gold-water nanofluid surface (B), where t_s is the time required for the surface temperature to reach the saturation temperature T_s^{100}, which is equal to 100°C, at atmospheric pressure.

Within the developed model, a diffusive mechanism of water vapor transport in air is adopted during the non-steady heating regime. As a result, changes in ambient humidity have little influence on the heating process of the volume of gold-water nanofluid. However, it is important to note that in practical applications, ambient humidity significantly affects the efficiency of converting solar radiation into thermal energy. This is because at high humidity levels, the intensity of the incident radiation flux onto the surface of the studied object will be noticeably reduced.

3.2 The efficiency of the evaporation process

The calculation results of the efficiency coefficient and mass flux density of water vapor in the steady-state evaporation process were determined based on the developed model and presented in Fig. 5. It should be noted that for each value of the working fluid layer thickness H, there exists an optimal nanoparticle concentration ϕ, and vice versa (Fig. 5B, D). Furthermore, as ϕ or H increases, the optimal value of the other parameter (i.e., H or ϕ) shifts towards its decrease. Therefore, for systems utilizing gold-water nanofluids as solar radiation absorbers, the optimal heating and evaporation regime lies within the region denoted as 1 in Fig. 5A and C.

It is important to note that in the process of steady-state vapor generation, the evaporation efficiency coefficient η_{evap} and the mass flux density j depend solely on the interfacial surface temperature T_s and the surrounding conditions, including temperature, humidity, and methods of vapor removal. Therefore, for effective vapor generation and obtaining clean water, it is advisable to develop a system with localized heating in the vicinity of the interfacial surface to reduce thermal losses associated with heating the liquid volume.

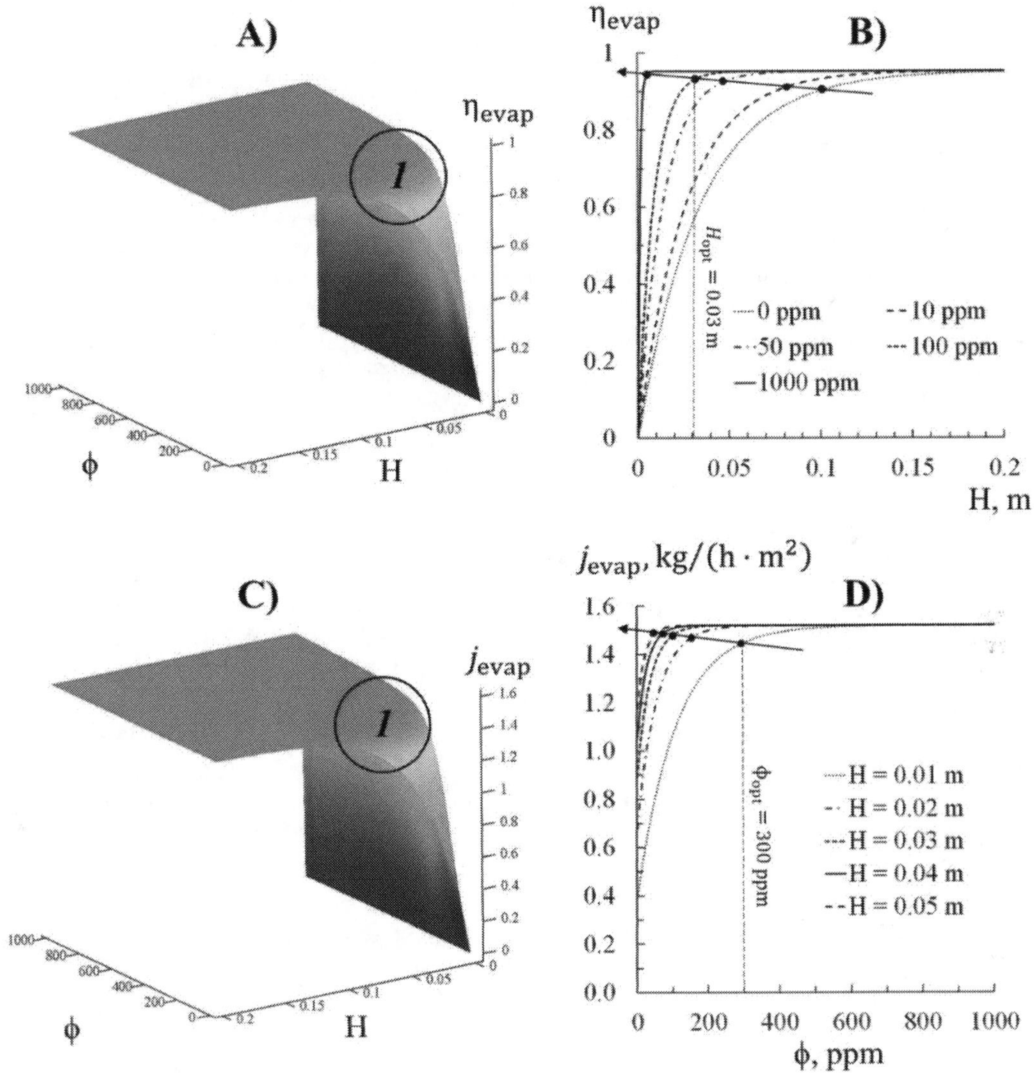

Fig. 5. The efficiency of evaporation (A, B) and vapor generation (C, D) in steady-state mode.

4 Conclusion

In this study, a mathematical model has been formulated to evaluate the efficiency of heating and evaporation in nanofluids. By conducting calculations specifically on gold-water nanofluid, several key conclusions can be drawn from the obtained results.

- For systems utilizing nanofluids as heat transfer fluids, the efficiency is influenced by parameters such as optical thickness (*H*), material type, nanoparticle concentration (ϕ), nanoparticle properties, and surrounding environmental conditions. It has been demonstrated that in the non-steady-state regime, the efficiency of such systems cannot be assessed solely based on the absolute value of the η_{heat} parameter, but rather through its time derivative. The results of the calculations conducted on the gold-water nanofluid have revealed the presence of optimal parameters (*H*, ϕ) at which the system can attain a steady-state regime within a specific timeframe and achieve a significantly elevated temperature level.

- In the context of solar-driven water purification systems, it is advantageous to strive for localized heating in the interfacial region to facilitate efficient vapor generation. In the case of utilizing nanofluids for vapor generation, optimal parameter ranges (H, ϕ) have also been demonstrated, as illustrated through calculations conducted on gold-water nanofluid.
- Although the developed model is simplified, it provides the capability to assess the efficiency of solar radiation conversion systems into thermal energy, considering multiple variables. Moreover, the findings of this study can serve as a fundamental basis for the advancement of more sophisticated models that incorporate the impact of additional factors, including complex geometries with forced fluid flow, the nanoparticle-base fluid interaction characteristics, solar radiation conversion mechanisms into thermal energy, and other pertinent variables.

The authors would like to acknowledge the financial support from the Russian Science Foundation, project 23-19-00840.

References

1. A.S. Dmitriev, A.V. Klimenko, Therm. Eng., **67**, 77–91 (2020)
2. A.S. Dmitriev, A.V. Klimenko, Therm. Eng., **70**, 551–572 (2023)
3. E.C. Okonkwo, I. Wole-Osho, I.W. Almanassra, Y.M. Abdullatif, T. Al-Ansari, J. Therm. Anal. Calorim., **145**, 1–56 (2020)
4. Srishti, Apurba Sinhamahapatra, Aditya Kumar, Energy Adv., **2**, 574–605 (2023)
5. M. Gao, C.K. Peh, H.T. Phan, L. Zhu, G.W. Ho, Adv. Energy Mater., **8**, 1800711 (2018)
6. H. Liu, Z. Huang, K. Liu, X. Hu, J. Zhou, Adv. Energy Mater., **9**, 1900310 (2019)
7. L. Zhou, X. Li, G.W. Ni, S. Zhu, J. Zhu, Natl. Sci. Rev., **6**, 562–578 (2019)
8. L. Cui, P. Zhang, Y. Xiao, Y. Liang, H. Liang, Z. Cheng, L. Qu, Advanced Materials, **30**, 1706805 (2018)
9. O. Mahian, E. Bellos, C.N. Markides, R.A. Taylor et al., Nano Energy, **86**, 106069 (2021)
10. S. Akilu, K.V. Sharma, A.T. Baheta, R. Mamat, Renew. Sustain. Energy Rev., **66**, 654–678 (2016)
11. Q.T. Tran, A.S. Dmitriev, Bulletin of MPEI, **6**, 110–119 (2023)

Comparative Assessment of Hydrothermal Gasification and Anaerobic Digestion using Aspen Plus and SuperPro

David Ballenghien[1], *Rajesh* S Kempegowda[2], *Wei-Hsin* Chen[3], and *Khanh-Quang* Tran[4*]

[1]374 water Inc, USA
[2]Inrigo As, Trondheim Norway.
[3]Department of Aeronautics and Astronautics, National Cheng Kung University, Tainan 701, Taiwan
[4]Department of Energy and Process Engineering, Norwegian University of Science and Technology, NO7491 Trondheim, Norway.

Abstract. Wet biomass is gaining increasing attention as an energy source globally. Various wet biomass materials like macroalgae, microalgae, sewage sludge, cattle manures, and food waste, are of high moisture contents, typically around 70% or more. There are two routes of high potential for converting these wet biomasss resoures to gaeous fuel, which are biomethanation and supercritical water gasification (SCWG). Biomethanation is conversion of wet-organic residues to biogas using via anaerobic digestion process. SCWG is a thermochemical conversion process, taking place in supercritical water producing enriched CH_4 and H_2 gases. In this study, the two processes are comperatively investigated using different process simulation softwares. Aspen Plus was employed for SCWG, while the biomethanation process was simulated with SuperPro. Various raw materials were used as input. We calculated the energy efficiency of the biomethanation process using experimental yields from the literature. For the SCWG process, we determined the higher heating value based on the simulated composition of the methane-enriched gas. Additionally, we conducted an economic analysis to compare the two processes, taking into consideration specific criteria relevant to the Norwegian context.

1 Introduction

There is serious speedup required in renewable energy particularly in biomass-based fuels. According to IEA data [1] "*global biofuel demand is projected to increase by 28% (41 billion liters) from 2021 to 2026*". This is mainly driven by post-Covid-19 recovery and government policies. As well, factors like transport fuel demand, costs, and policy design will influence growth patterns. Asia is expected to surpass Europe in biofuel production. There is a tendency of another oil-embargo expected in coming years. These uncertainties tend government responses to feedstock prices, which may impact the demand. There are serious discussions

* Corresponding author: khanh-quang.tran@ntnu.no

in major economies like the United States, Europe, India, and China could potentially push for doubling the production of biofuels.

However, achieving this potential presents significant challenges related to sustainability and cost. In the context of Norway, characterized by its cold climate, experiences slower biomass growth compared to tropical regions. In contrast, macroalgae, which thrives in Norwegian waters, holds great promise as a biofuel source. By dedicating Norwegian sea area of 1% [2], report estimated that 60% of the country's fuel needs could be met by biofuel produced. Another report estimated that an area equivalent to Norway's salmon production zones (800 km^2) could sustain the production of 470,000 tonnes of ethanol which can cut down a significant portion of the country's gasoline and diesel consumption [3]. Considerable research efforts are already underway, in Norway and internationally, to cultivate macroalgae for biofuels. Notable national initiatives include Macrobiomass, the Norwegian Center for Seaweed Technology, and projects by Seaweed Energy Solution and Statoil.

Lignocellulosic residues from existing processes, such as fermentation residues from lignocellulosic ethanol and biogas plants, represent another potential biomass feedstock to meet future biofuel needs in Norway. This alternative biofuels production pathway is instrumental in enabling Norway to develop cost-competitive biofuel value chains and achieve ambitious targets set by national and international policies and strategies. This article aims to compare two processes for converting wet biomass to gaseous biofuels: biomethanation and supercritical water gasification (SCWG). The goal is to establish reliable figures describing energy efficiency for different types of feedstock based on process simulation and experimental data from the literature.

Wet biomass, which does not typically compete with food production, includes feedstocks such as macro- and microalgae, wastewater-treatment-plant sludge, cattle manure, and high-moisture content waste from the food industry. These feedstock, often waste streams or by-products, can lead to significant pollution if mishandled due to their high organic contents. Current treatment methods involve biomethanation or aerobic digestion processes, but these processes have limitations, including the inability to digest lignin-rich materials and the challenges of storing the resulting digestate sludges, especially in densely populated areas.

SCWG is pushed as a solution to these issues. It achieves nearly complete conversion on a pilot plant scale, leaving virtually no organic matter after the reactor. SCWG offers several advantages over biomethanation process, including higher throughputs, energy efficiency, the ability to handle mixed feedstocks, and no need for specialized microbial cultures or enzymes.

In contrast to biomethanation, SCWG requires harsh conditions, including high pressure and temperature (above 22.1 MPa and 374°C). Heat losses at such temperatures are higher, and substantial energy is needed to reach the required pressure. While SCWG claims higher overall energy efficiency than biomethanation, when considering the investment, operation, and maintenance costs, the preference between the two processes is not straightforward. To help in determining the best process design, a detailed Pinch analysis is performed to optimize heat integration and implement heat exchangers in the SCWG process. This analysis evaluates the overall energy efficiency of methane-rich gas production. Additionally, the SCWG process is compared with conventional anaerobic processes in terms of technical and economic viability under Norwegian conditions.

2 Process description of two process under compare

2.1 Hydrothermal gasification

Fig. 1 depicts the SCWG process, accommodating a biomass input rate of 20 tons per hour with an initial moisture content of 80%. The process initiates with the ambient-temperature biomass and undergoes a series of pre-treatment steps, including chipping, milling, weak acid hydrolysis to reduce alkalinity, and coarse filtration based on the characteristics of the feedstock. Subsequently, the high-pressure pump elevates the pressure to 34.5 MPa, while the temperature is increased to approximately 377°C. This temperature increase is achieved through a two-step process: first, via heat exchange, and second, by an additional heat supply from a combustor fuelled by a fraction of the product gas. The final temperature is selected to align with the desired gasifier outlet temperature of 600°C.

After leaving the gasifier, the ash and supercritical mixture are separated. The ash is removed from the process, while the supercritical stream passes through multiple heat exchangers to preheat the incoming stream. It then undergoes the first flashing tank, maintaining a pressure of 34.5 MPa. The gas stream is enriched with hydrogen, with a hydrogen content as high as 95%, whereas the liquid stream is extracted and routed through a second flashing tank, where the pressure is finally reduced to 1.0 atm. This step aims to recover gases dissolved in the water. A portion of this stream is directed to a combustor, which supplies the additional heat needed in the preheating section, especially when heat exchange alone cannot suffice. The remainder of the stream is available for further use, similar to the hydrogen-enriched gas stream. The liquid stream following the second flashing tank is removed from the process. It predominantly consists of water, with some inorganic and unconverted organic compounds.

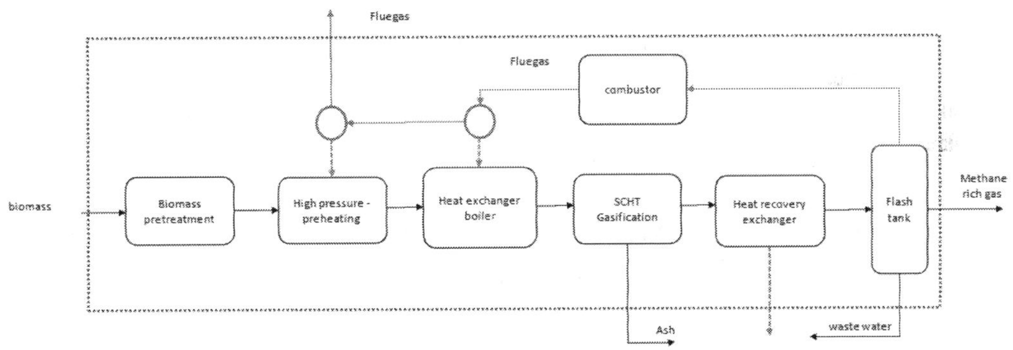

Fig. 1. Process configuration of wet-biomass processing using hydrothermal gasification under super-critical conditions for CHP

2.2 Anaerobic digestion

Biomethanation is an anaerobic process involving four steps: hydrolysis, acidogenesis, acetogenesis, and methanation. Biomethanation is applied in various wastewater, slurry, and solid waste treatment processes, utilizing different reactor types and conditions to maximize energy output and process efficiency. In this example, data gathered from Inrigo Biosystem data for three-stage anaerobic reactor after slurry pre-treatment are proposed to enhance the biomethane yield. Stage thermal hydrolysis after slurry prepared by milling, pumped to thermal hydrolysis which take place at 70°C to meet the PAS rules as per pasteurization concern. Then pasteurized slurry supplied to anaerobic digester and temperature of the slurry cooled down by mixing with recycled sludge. The main criteria to avoid inhibition is by controlling total organic nitrogen below 5%.

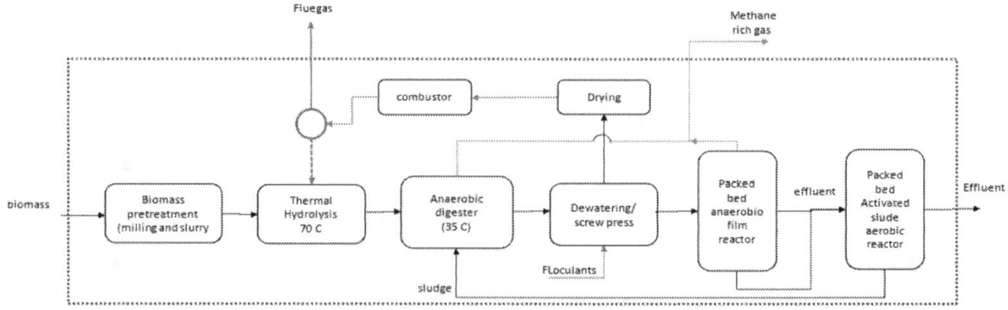

Fig. 2. Block flow diagram of biomethanation process.

Fig. 2 depicts the outline of the biomethanation process. The pretreatment involves chopping and milling waste feedstocks. These pretreated materials are then heated to 70°C during thermal hydrolysis and subsequently transferred to a mesophilic anaerobic reactor maintained at 35°C. The hydraulic retention time for anaerobic digestion considered here ranges from 15 to 20 days, resulting in the production of a CH_4 and CO_2 mixture. The typical conversion of chemical oxygen demand (COD) in the reactor is approximately 60%. Unconverted effluents from the first-stage anaerobic digester are directed to the mechanical dewatering system, known as the "mechanical dewatering system," where the cake can yield around 30% dry mass. Further drying is carried out using a flue gas dryer. The dried biosolid sludge is incinerated in the boiler/combustor to provide heat for the preheating section.The effluent from the reactor is then sent to an anaerobic biofilm reactor to reduce unconverted COD and enhance biomethane production. Subsequent effluent can undergo aerobic treatment in a packed bed/trickling bed aerobic reactor to reduce COD and meet water recycling objectives. The biomass input is adjusted to 20 tons per hour with a moisture content of 10%, ensuring that an equivalent amount of dry matter is utilized in the SCWG process, enabling a fair and meaningful comparison.

3 Methodology of process system design

Table 1 Proximate and ultimate analysis of the different feedstock's (weight % in dry basis) and ash analysis for Kelp

Proximate	Kelp	Birch	Water Hyacinth
Moisture	0.00	0.00	0.0
Ash	45.63	0.22	22.1
Volatile	49.36	89.43	62.0
Fixed Carbon	5.01	10.35	15.90
Ultimate			
Carbon	24.19	48.62	36.4
Hydrogen	3.08	6.34	4.77
Nitrogen	2.62	0.09	1.82
Sulphur	0.70	0.05	0.4
Ash	30.00	0.022	22.1
Oxygen	23.78	44.90	32.6
Chlorine	15.63	0.00	1.9
HHV (MJ/kg)	9.33	19.80	14.81

 https://doi.org/10.1051/e3sconf/202451403003

Fig. 3 illustrates the comprehensive process simulation of SCWG using Aspen Plus. However, for the biomethanation process, the requirement for well-established and validated data on anaerobic digestion necessitates the utilization of Superpro Designer [5], a bioprocess simulation software. The methane yield and, consequently, the efficiency of the biomethanation process are determined for various feedstocks based on experimental data sourced from the literature. Additionally, a heat integration strategy were based on the Aspen Plus simulation screened temperatures. Then, the techno-economic viability of the two process were compared based on an annuity analysis. The attributes of the feedstocks examined in the context of Norwegian conditions are presented in Table 1.

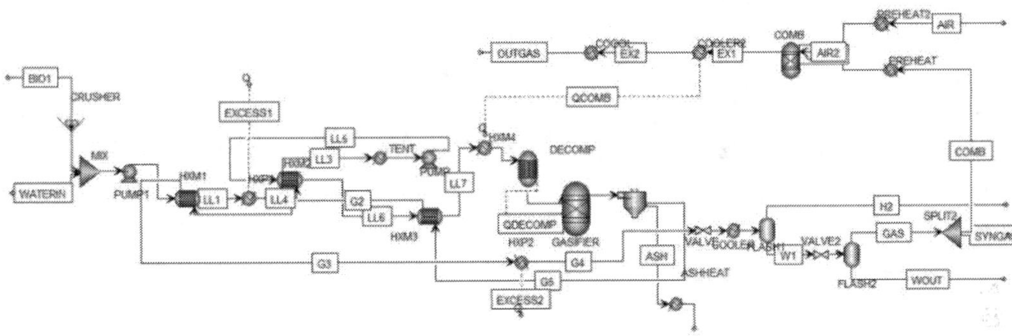

Fig. 3. SCWG process simulation in Aspen Plus

Based on the process flow diagram shown in Fig. 3, mass and energy balances for each process and unit were performed using the Aspen Plus. This package comprises numerous sub-models that simulate individual units and components. Calculations were conducted for each of the sub-models to determine thermodynamic properties and describe the behaviour of every single component and mixture. Equations and data required for the calculations were obtained from Aspen Plus databases. Input data were gathered from various sources, including experimental and literature data. Due to the high pressure and temperature of the process, the behavior of compounds deviates significantly from ideality. Typically, for petrochemical reactions, Peng-Robinson or RK-Soave equations are employed [4–6] . In our case, we adopted the Peng-Robinson equation, which is widely accepted for solids handling.

3.1 Hydrothermal gasifier modelling in Aspen Plus

The synthesis of the base configuration flowsheet is developed using Aspen Plus, as illustrated in Fig. 3. In Aspen Plus, the defining of properties for biomass is done according to Aspen Plus solid modelling methodology. Gasifier modelled by combination of decomposition reactor R-Yield where biomass decomposed in to its basic proximate and ultimate fractional yields. Carbon is transformed into solid carbon, oxygen into O_2, nitrogen into N_2, hydrogen into H_2, sulfur into solid sulfur (S), and chlorine into Cl_2. Water remains as water, and ash remains as ash, with ash being a pre-tabulated compound in Aspen Plus. Gasification reactor modelled based on non-stoichiometric model by the well-known Gibbs approach. It is worth noting that ash is assumed not to undergo any chemical interactions within the process, which is a significant simplification. It is also important to mention that the catalytic effects of mineral compounds, such as KOH, Na_2CO_3, and K_2CO_3, have been reported [7, 8] and leaching and demineralization effect of inorganics ignored in the mass and energy balance. These compounds can readily form from ash in supercritical water, especially since the ion dissociation constant (K_w) of water decreases significantly after the critical point, leading to the precipitation of most ionic compounds [9]. However, a comprehensive modelling effort would be necessary to fully understand how all components of ash interact with other compounds, but this level of detail falls outside the scope of this work. The gasifier is represented in the model by a Gibbs reactor.

3.2 Gas upgrading and separation process

K-values predicted with the thermodynamics property Peng-Robinson were used to calculate the mole fraction of each chemical in the vapour and liquid phase shown below equation 3.1.

$$y_i = K_i x_i \tag{3.1}$$

Where y_i is the molar fraction of compound i in the gas phase, and x_i the molar fraction of the compound in the liquid phase.

Boiler: The boiler is simulated as a Gibbs reactor, aiming to minimize the Gibbs free energy of the mixture in the presence of air (with oxygen at a 1:2 molar ratio). The reactions for boiler combustion are shown below:

$$H_2 + \tfrac{1}{2}O_2 \rightarrow H_2O \tag{3.2}$$

$$CO + \tfrac{1}{2}O_2 \rightarrow CO_2 \tag{3.3}$$

$$CH_4 + 2O_2 \rightarrow 2H_2O + CO_2 \tag{3.4}$$

It receives a fraction of the generated gas to meet the preheating requirements. The hot exhaust gas aids in the final preheating step before the gasifier through heat exchange with the incoming stream. The usable temperature range is limited to around 400°C, given that initial preheating steps rely on heat recovery from the outgoing gas.

3.3 Simulation of the biomethanation process

This section provides details on the simulation of the biomethanation process using the commercial software Superpro Designer. The simulated sub-processes include anaerobic digestion, aerobic reactor, mechanical dewatering, and sludge combustion for steam generation, as illustrated Fig. 4. The mass yields of each compound depend on the characteristics of the biomass feedstock type. Experimental yields validated under the anaerobic reactor model of Superpro for is valid only for the model compound of Water Hyacinth $C_6H_{12}O_{6.8}$ as methane 14.8%, CO2 40.8%, sludge 34.6%, and lignin 10%. Volatile Solid (VS) is the fraction of total solid that has been volatilized after combustion at 550°C.

Aeróbic Reactor: The reactions in the aerobic reactor, designed to treat water by oxidizing the remaining organic compounds, primarily produce CO_2. Some ammonium radicles may also be released based on substrate N_2 concentration, which is subsequently oxidized into nitrates. The denitrification reaction has not been included here, as our focus is on the energy balance of the process. Methane produced from the anaerobic reactor and the organic fraction of the sludge are considered energy products. To prevent the excessive accumulation of sludge, it is further sent to the belt filtration system.

Belt Filtration: The sludge water content after belt filtration is set at 70% [10]. Mechanical water removal can only concentrate the sludge to around 30% solid content. However, for it to be usable as a fuel, the sludge needs to be dried to a moisture content of around 25%. Therefore, for every 1 kg of sludge, approximately 2.0 kg of water must be thermally removed, consuming 4.9 MJ of energy. The dried sludge is then fed into the boiler process for heat production.

https://doi.org/10.1051/e3sconf/202451403003

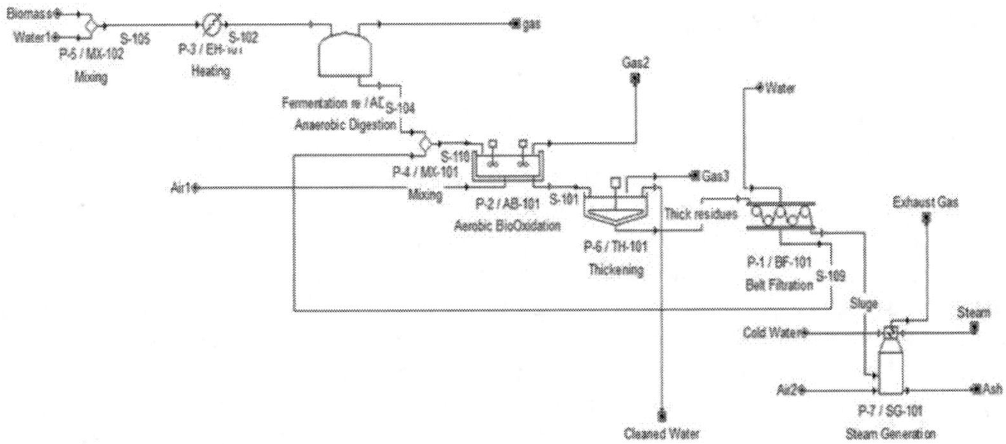

Fig. 4. Biomethanation process designed in Superpro

4 Results and discussions

4.1 Sensitivity analysis

Sensitive analysis were carried to know the influence of gasification temperature on the composition of the product gas, of which the results are presented in the Fig. 5. This sensitivity analysis shows that H_2 is favoured at high temperature, whereas CH_4 gets under 10% molar fraction beyond 1000 K or 727°C. The flashing pressure sensitivity analysis shows that the composition of the flows can be controlled through a pressure control in the flashing tanks. Methane molar fraction is the highest around 10 bar, whereas H_2 keeps increasing with the pressure. For different types of feedstock, the yields were obtained from experimental data found in the literature[11,12] and supported by in-house industrial data from Inrigo.

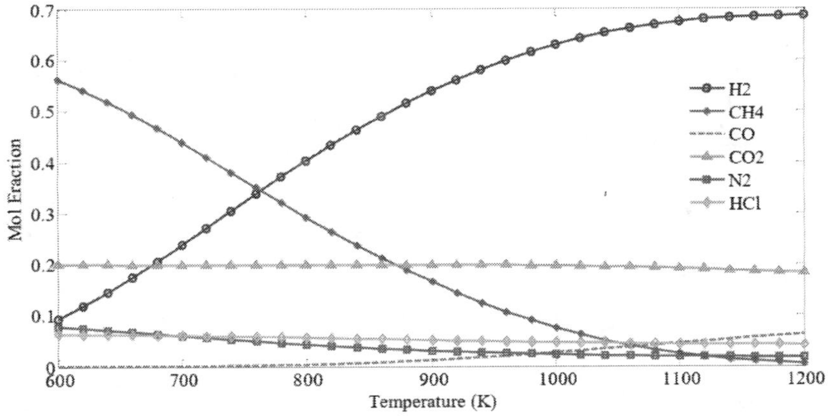

Fig. 5. Influence of the gasification temperature on the composition of the produced gas

 https://doi.org/10.1051/e3sconf/202451403003

4.2 Heat recovery process and design modification

The heat integration is essential in the SCWG process and impacts dramatically the energy efficiency. The methodology is detailed in [8]. The high temperature and pressure of the process and the large quantity of water in biomass makes it certain challenges with selection of heat exchanger due to corrosion after the gasifier. In our case where the purpose is to use wet biomass, recycling of water is not interesting since the biomass is contained it and there is already an excess of water. However, if SCWG is to be applied to dry feedstock such as straw or wood, a stream of high pressure and high temperature water is mixed with the stream of biomass. Most of the previous studies in that field based their efficiency analysis on assumptions of the heat exchange efficiency, or energy recovery efficiency. Pinch analysis was carried out to propose a relevant heat integration and implementation of heat exchangers. Due to uncommon conditions of temperature and pressure, an average value of the thermal capacity C_p of the different fluids would be irrelevant, therefore the specific enthalpy of each stream has been calculated numerically by steps of 0.5 K with Aspen Plus, and average values of C_p have been calculated on the same ranges of temperature. Fig. 6 shows the heat integration strategies for hydrothermal gasifier.

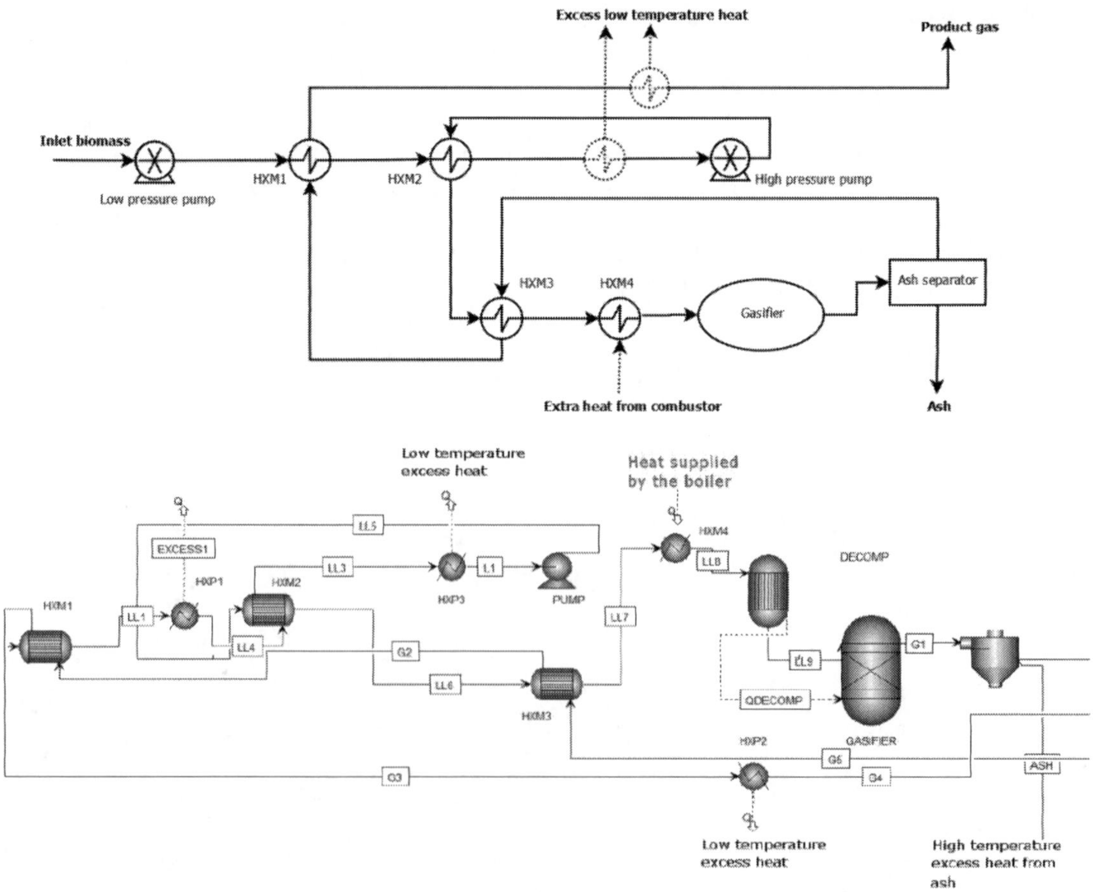

Fig. 6. (a) Heat integration strategy and (b) implemented in Aspen plus

4.3 Composite diagrams for different types of feedstock

Fig. **7** shows the composite diagrams of the flows to be cooled (upper curve) and those to be heated. The impact of the minimum temperature approach on the quantity of heat that has to be supplied by the external source with minimum temperature approach of 15 K and 30 K are illustrated. It is clear from Fig. 7 that the available energy from the hot streams is larger than the energy requirement to heat the cold streams. For instance for Kelp, in the formerly given conditions, the total enthalpy rate available at the hot streams is 9847 kJ/s, which needs to be given away, whereas the enthalpy rate required at the cold streams is 9622 kJ/s. However, for a minimum temperature difference of 30 K in the heat exchangers, part of the heat must be supplied by an external source due to insufficient high temperature heat. This amount of heat depends of course on the temperature difference, which will influence both the heating rate and the size of the heat exchangers, and therefore the investment costs. The same pattern occurs with Birch and Kelp as shown in figure. From an energy efficiency point of view, efficient heat exchangers should be preferred however, the larger investment cost may hinder the solution.

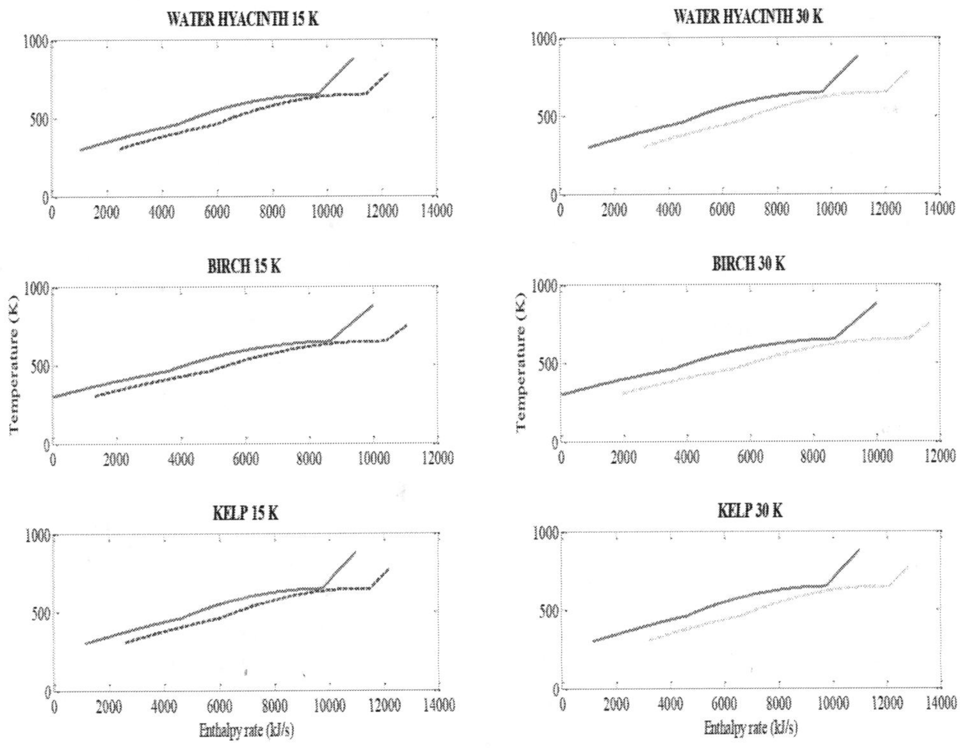

Fig. 7. Composite diagrams for WATER HYACINTH, BIRCH and KELP at 15K and 30K

4.4 *Energy efficiency of SCWG*

The results of the simulation are given in

Table 2, consisting of syngas composition and energy efficiency for different types of feedstocks with two different heat integration strategies.

Table 2 Energy efficiency and composition of the produced gas with SCWG for different fuels.

Feedstock	**KELP**	**Birch**	**Water Hyacinth**	**Literature (1)**
Component	**Mol Fraction**			
H_2	0.31	0.24	0.28	0.29
CH_4	0.18	0.33	0.29	0.25
CO	0.01	0.01	0.01	0.01
CO_2	0.35	0.41	0.39	0.43
N_2	0.02	0.00	0.01	0.03
SO_2	0.01	0.00	0.00	0.00
HCl	0.12	0.00	0.01	0.00
Temperature approach	**ΔT_{min} [K] = 30 K**			
Energy efficiency η	55% [a]	77%	70%	64.8% [b]
Fraction Methane-enriched gas sent to boiler	0.54	0.24	0.34	-
H2 rich flow [kg/h]	41.5	64.5	55.9	-
Average HHV [MJ/kg]	90	83	85.8	-
Output Methane-enriched gas sent to boiler [kg/h]	894	1855	1248	-
Average HHV [MJ/kg]	7.9	13.9	13.1	-
Temperature approach	**ΔT_{min} [K] = 15 K**			
Energy efficiency η	69% [c]	84%	80%	-
Fraction Methane-enriched gas sent to boiler	0.36	0.15	0.22	-
Boiler after HX temperature [K]	680	680	680	-

4.5 Anaerobic digestion process

Yields are taken from experimental data found in the literature [9] for different types of feedstock. They are usually expressed as Nm^3 methane per kg of volatile solid (VS). The methane yields and the simulated energy efficiency are tabulated shown below as a comparison to hydrothermal gasifier-based methane rich gas production as shown in Table **4**.

Table 3 Biomethanation energy efficiency for different feedstocks (Anaerobic digestion)

Feedstock	**Water Hyacinth , (theortical)**	**Water Hyacint h**	**Water Hyacinth Pretreatment**	**Laminaria Hyperborea**	**Kelp**
Literature	[13]	[11]	[11]	[11]	[11]
Water [kg]	9000	9000	9000	9000	9000
Biomass [kg]	1000	1000	1000	1000	1000
VS[a] added [kg]	1000	779	779	820	550
Ash content [wt %]	0	22.1	22.1	18	45
VS reduction	0.554	0.498	-	-	0.508
CH_4 produced [kg]	148	92	155	131	96
HHV Biogas [MJ]	8234	5145	8639	7300	5315
Preheating duty [MJ]	680	680	680	680	680
Evaporation of water in sludge	3499	2549	-	-	4119
Heat loss from the reactor [MJ]	20	13	13	17	14
Electric demand [MJ]	520	520	520	520	520
Dry sludge produced [kg]	446	612	-	-	721

Ash content of dry sludge [wt%]	0.00%	36.11%	-	-	62.50
Dry, ash free sludge HHV [MJ/kg]	5.99	20.71	-	-	12.45
Energy content in sludge [MJ][b]	2669	8097	-	-	3364
Temperature [°C]	45	35	35	35	35
Residence time [day]	30	15	-	24	18
Size of the anaerobic reactor [m^3]	302	150	-	238	177
Working / vessel volume	90%	90%	-	90%	90%
Energy input [HHV, MJ]	11420	14810	14810	16540	9330
Efficiency with heat recovery	68%	31%	55%	41%	51%
Excess heat from the boiler [MJ]	-580	3906	-	-	1450

4.6 Techno-Economic analysis

Cost estimations of equipment are based on the principles engineering design involving various process design principles and cost of equipment modified from various literatures [13,14]. A scaling factor of 0.6 was applied to SCWG plant costs, and the same scaling factor was used to assess biomethanation plant costs (CAPEX). Direct and indirect costs were drawn from Peter et.al [15], Gasif et.al [14], and our previous publications [16–18]. Costs are expressed in US dollars, and both capital expenditures (CAPEX) and operating expenditures (OPEX) were provided for both SCWG and Biomethanation plants. Cost modelling approach were well documented by us in various sponsored research projects and publications [16–20]. Estimated annualized cost for SCWG and biomethanation process are shown in Table .

Table 4 Summary of the annuitized cost

	SCWG process Million $	Biomethanation process Plant in Million $
Annual cost of capital	4.02	2.35
Annual operation and maintenance cost	6.84	3.86
TOTAL	10.86	6.22

The comparison involves plotting, as shown in Fig. 8, the cost of producing product gases ($/GJ) against the biomass purchase cost. We conducted this comparison without considering combined heat and power (CHP) for a fair assessment. Different biomass purchase prices were considered, ranging from $20/ton (the current price for macroalgae in the Norwegian alginate industry) to as low as negative prices, like -60$/ton, often applicable to materials such as sewage sludge, where disposal costs can be significant. The results show that SCWG generally incurs higher production costs due to its greater investment expenses. However, the choice of feedstock significantly influences production costs. Notably, SCWG competes effectively with biomethanation when a scaling factor of 0.6 is applied for water hyacinth feedstock across a wide range of purchase prices. Conversely, for Kelp, SCWG consistently yields higher production costs compared to biomethanation, regardless of the scaling factor.

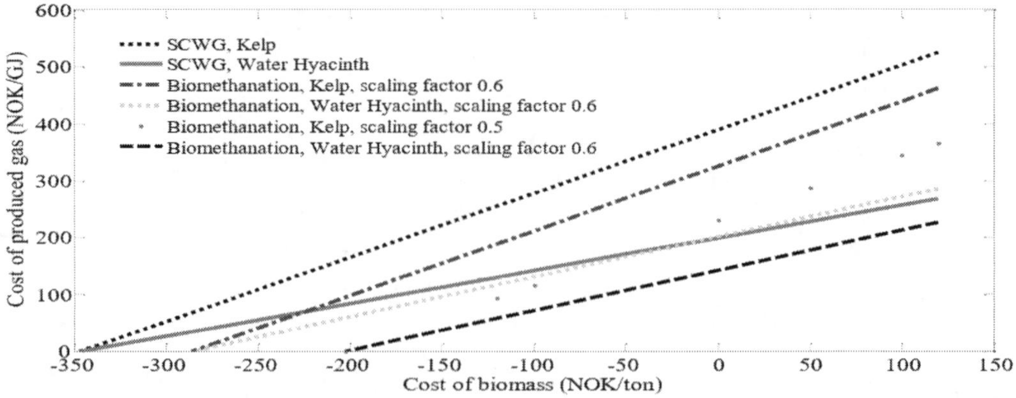

Fig. 8. Influence of the purchase cost of biomass on two feedstocks for SCWG and Biomethanation, at 0.5 and 0.6 scaling factor.

5 Conclusions

In this study, we have examined both biomethanation and supercritical water gasification processes with a focus on energy efficiency, heat integration and cost-economics. We employed two different minimal temperature approaches for the SCWG process, namely 30°C and 15°C, which reflect the performance of heat exchangers and have a significant impact on overall energy efficiency. At a 30°C minimal temperature approach, the energy efficiency for Kelp in the SCWG plant was 55%, which increased to 69% at a 15°C minimum temperature approach, while biomethanation achieved 51% energy efficiency. This difference becomes even more pronounced with feedstocks containing lignin, where biomethanation struggles without pretreatment. For water hyacinth, the SCWG plant achieved 70% energy efficiency at a 30°C minimal temperature approach, while biomethanation with steam pretreatment reached 55% and without pretreatment, only 31%. The key advantage of SCWG is its exceptionally high energy efficiency, particularly for feedstocks containing lignin when compared to biomethanation. Additionally, SCWG offers near-total biomass conversion and faster operations within a smaller footprint.

Acknowledgements

This publication has received funding from the Norway Grants 2014–2021 via the National Center for Research and Development. The article has been prepared within the frame of the project: "Negative CO_2 emission gas power plant" - NOR/POLNORCCS/NEGATIVE–CO_2–PP/0009/2019–00 which is co-financed by programme "Applied research" under the Norwegian Financial Mechanisms 2014–2021 POLNOR CCS 2019 - Development of CO_2 capture solutions integrated in power and industry processes.

References

1. IEA. Renewables 2021, IEA, Paris 2021.
2. Fernand F, Israel A, Skjermo J, Wichard T, Timmermans KR, Golberg A. Offshore macroalgae biomass for bioenergy production: Environmental aspects, technological achievements and challenges. Renewable and Sustainable Energy Reviews 2017;75:35–45. https://doi.org/https://doi.org/10.1016/j.rser.2016.10.046.

3. Skjermo J, Aasen IM, Arff J, Broch OJ, Carvajal AK, Christie H, et al. A new Norwegian bioeconomy based on cultivation and processing of seaweeds: Opportunities and R&D needs, 2014.
4. Recalde M, Amladi A, Venkataraman V, Woudstra T, Aravind PV. Thermodynamic analysis of supercritical water gasification combined with a reversible solid oxide cell. Energy Convers Manag 2022;270:116208. https://doi.org/https://doi.org/10.1016/j.enconman.2022.116208.
5. Macrì D, Catizzone E, Molino A, Migliori M. Supercritical water gasification of biomass and agro-food residues: Energy assessment from modelling approach. Renew Energy 2020;150:624–36. https://doi.org/https://doi.org/10.1016/j.renene.2019.12.147.
6. Yan Q, Guo L, Lu Y. Thermodynamic analysis of hydrogen production from biomass gasification in supercritical water. Energy Convers Manag 2006;47:1515–28. https://doi.org/10.1016/j.enconman.2005.08.004.
7. Boukis N, Stoll IK. Gasification of biomass in supercritical water, challenges for the process design—lessons learned from the operation experience of the first dedicated pilot plant. Processes 2021;9:1–17. https://doi.org/10.3390/pr9030455.
8. Yanik J, Ebale S, Kruse A, Saglam M, Yüksel M. Biomass gasification in supercritical water: II. Effect of catalyst. Int J Hydrogen Energy 2008;33:4520–6. https://doi.org/10.1016/j.ijhydene.2008.06.024.
9. Peterson AA, Vogel F, Lachance RP, Fröling M, Antal MJ, Tester JW. Thermochemical biofuel production in hydrothermal media: A review of sub- and supercritical water technologies. Energy Environ Sci 2008;1:32–65.
10. Olivier J, Vaxelaire J. The Prediction of Filter Belt Press Dewatering Efficiency for Activated Sludge By Experimentation on Filtration Compression Cells. Environ Technol 2005;25:1423–30. https://doi.org/10.1080/09593332508618474.
11. Nallathambi Gunaseelan V. Anaerobic digestion of biomass for methane production: A review. Biomass Bioenergy 1997;13:83–114. https://doi.org/https://doi.org/10.1016/S0961-9534(97)00020-2.
12. Kunatsa T, Xia X. A review on anaerobic digestion with focus on the role of biomass co-digestion, modelling and optimisation on biogas production and enhancement. Bioresour Technol 2021;344:126311. https://doi.org/10.1016/j.biortech.2021.126311.
13. Matsumura Y. Evaluation of supercritical water gasification and biomethanation for wet biomass utilization in Japan. Energy Conversion and Management - ENERG CONV MANAGE 2002;43. https://doi.org/10.1016/S0196-8904(02)00016-X.
14. Gasafi E, Reinecke M-Y, Kruse A, Schebek L. Economic analysis of sewage sludge gasification in supercritical water for hydrogen production. Biomass Bioenergy 2008;32:1085–96. https://doi.org/https://doi.org/10.1016/j.biombioe.2008.02.021.
15. Peters MS, TKD, & WRE. Plant design and economics for chemical engineers (5th ed.). 5th ed. 2003.
16. S Kempegowda R, Skreiberg Ø, Tran K-Q. Cost modeling approach and economic analysis of biomass gasification integrated solid oxide fuel cell systems. Journal of Renewable and Sustainable Energy 2012;4. https://doi.org/10.1063/1.4737920.
17. Del Alamo G, Kempegowda RS, Skreiberg Ø, Khalil R. Decentralized Production of Fischer–Tropsch Biocrude via Coprocessing of Woody Biomass and Wet Organic Waste in Entrained Flow Gasification: Techno-Economic Analysis. Energy & Fuels 2017;31:6089–108. https://doi.org/10.1021/acs.energyfuels.7b00273.

18. Kempegowda RS, Skreiberg Ø, Tran K-Q, Selvam PVP. Techno-economic Assessment of Thermal Co-pretreatment and Co-digestion of Food Wastes and Sewage Sludge for Heat, Power and Biochar Production. Energy Procedia 2017;105:1737–42. https://doi.org/https://doi.org/10.1016/j.egypro.2017.03.498.
19. Kempegowda RS, del Alamo Serrano G, Güell BM, Tran K-Q. Techno-economic Analysis of Biomass to Fischer-tropsch Diesel Production with and without CCS Under Norwegian Conditions. Energy Procedia 2014;61:1248–51. https://doi.org/https://doi.org/10.1016/j.egypro.2014.11.1070.
20. Kempegowda RS, Tran K-Q, Skreiberg Ø. Techno-economic assessment of integrated hydrochar and high-grade activated carbon production for electricity generation and storage. Energy Procedia 2017;120:341–8. https://doi.org/https://doi.org/10.1016/j.egypro.2017.07.223.

Optimizing NACA2414 airfoil aerodynamics with PARSEC parametrization and Genetic Algorithm

Anupam Krishnan[1*], *Abdulkareem* Sh. Mahdi Al-Obaidi[1], and *Lee* Ching Hao[1]

[1]School of Engineering, Faculty of Innovation and Technology, Taylor's University, Taylor's Lakeside Campus, Subang Jaya, Selangor DE, Malaysia

Abstract. This research endeavours to contribute to the broader field of wind turbine aerodynamics by investigating a method to enhance the performance of airfoils, with specific attention to the NACA2414 airfoil at a 15-degree angle of attack. The study explores the airfoil's performance across a range of -10 to 15 degrees angle of attack. It employs both PARSEC parametrization and Genetic Algorithm optimization, achieving significant advancements. At a 15-degree angle of attack, post-optimization, the lift coefficient for the NACA2414 airfoil exhibits a remarkable increase to 1.5275 at a Reynolds number of 10^5, surpassing the original airfoil's performance of 1.2407. This progress highlights the effectiveness of utilizing PARSEC parametrization and Genetic Algorithm optimization, particularly in low-speed wind turbine applications. While emphasizing potential applications in low-wind-speed scenarios, the findings underscore the significance of these techniques in renewable energy production.

1 Introduction

In the pursuit of sustainable energy, wind turbines stand as a crucial player in the realm of renewable resources. These towering structures capitalize on the aerodynamic force contained in the wind to generate electricity, offering an eco-friendly alternative to traditional methods. Particularly vital in low wind speeds, even slight improvements in aerodynamic performance significantly impact power production. This emphasis has fuelled a surge in research on airfoil optimization, striving to enhance the efficiency of wind turbines.

Recent studies have explored innovative approaches to airfoil optimization. Researchers focused on enhancing the performance of a NACA0012 airfoil, tackling uncertain flow conditions with the Taguchi method [1]. Notably, the modified airfoil not only improved average performance but also reduced variability under uncertain conditions, showcasing an impressive 69% overall robustness improvement. In parallel, the introduction of Bézier-Generative Adversarial Network (Bézier-GAN) in machine learning demonstrated accelerated convergence and outperformance of other parameterization methods, highlighting the role of learned geometric parameterizations in complementing traditional methods [2].

*Corresponding author: anupam.krishnan@sd.taylors.edu.my

The Design, Network, Optimization methodology, encompassing geometric shape functions and optimization algorithms, introduced PARSEC parametrization and a Particle Swarm Optimization model. This approach showed potential for handling multimodal problems and ensuring consistent results, regardless of initialization [3]. Tortora *et al.* [4] sought to develop an innovative parameterization and fitting approach, dividing airfoils into three sections: central box, leading edge, and trailing edge. This method ensured smooth continuity between adjacent segments, validated through geometric and aerodynamic property comparisons.

In the quest for airfoil optimization, Mukesh *et al.* [5] employed the PARSEC parameterization method to analyse the geometric properties of the NACA2411 airfoil. Using Genetic Algorithm (GA), they successfully improved the wing's performance, validated through tests in a subsonic wind tunnel. Notably, the study achieved a remarkable closeness between estimated and empirical lift coefficient values, with a minimal 4% error. Extending the optimization horizon, the combined use of Bezier and PARSEC parameterization approaches emerged as a versatile method applicable to a diverse range of airfoil shapes [6]. Building on this, Krishnan *et al.* [7] conducted a comprehensive review exploring optimization methods applicable to both wind turbines and airfoils. The focus zeroed in on the efficacy of methods such as GAs and multi-parameter tuning in yielding optimal results. Transitioning to specific airfoil optimization strategies, a study concentrated on the NACA4415 airfoil, employing the Bézier curve for parameterization and GA for optimization [8]. This research primarily concentrated on single-objective optimization, particularly enhancing the lift coefficient for specific AoAs.

Shifting to supersonic airfoils, another study leveraged GA for the selection and optimization with the overarching goal of maximizing efficiency at a specific lift coefficient [9]. The study explored the impact of design variables on airfoil performance across various Mach numbers and lift coefficients, underscoring the pivotal role of effective parameterization in achieving improved designs with expanded operational capabilities. Further innovations emerged in a recent study that introduced an indoor evolutionary algorithm for single-objective optimization [10]. This study, focusing on enhancing lift coefficients, employed GAs, and utilized both polynomial-based and radial basis function methods to parameterize the NACA2412 airfoil. Notably, the PARSEC method outperformed the latter method, showcasing improved lift coefficients and stall region preservation. In summary, these studies collectively showcased a spectrum of approaches in airfoil optimization, seamlessly transitioning from general methods to specific applications and methodologies.

Expanding on advanced airfoil designs, morphing airfoils emerge as a cutting-edge solution, capable of dynamically adjusting their geometry to optimize performance in varying aerodynamic conditions. These airfoils possess the capacity to adjust key aspects of their geometry, including chord length, camber, and twist, with the aim of enhancing performance and efficiency across various conditions. Proper techniques to parametrise an airfoil and the implementation of evolutionary algorithms are the most suitable methods to achieve higher aerodynamic efficiency in unfavourable conditions. A 13% increase in the maximum lift coefficient for a NACA0012 airfoil was observed at low Reynolds number (Re) for a morphed trailing edge when compared to a hinged flap [11]. The tip deflection angle and the length were found to be critical as per the study. The effects of morphing blades on pressure distribution, airflow speed, and discharge patterns were studied, revealing an improvement in power generation capabilities of Vertical Axis Wind Turbines [12-13]. Researchers examined the influence of trailing edge thickness and its morphing on a NACA0012 airfoil by conducting a 2D Computational Fluid Dynamics analysis using the k-ω Shear Stress Transport model to study the airflow. The analysis revealed that as the angle of inclination increased, the lift to drag ratio also increased [14]. Previous studies have shown that enlarging the trailing edge thickness up to 0.7% can have positive impact on reducing the aerodynamic drag of an airfoil [15].

In this unique study, we apply PARSEC parametrization and GA optimization for the NACA2414 airfoil, prioritizing the lift-to-drag ratio over singular parameters like the lift coefficient. This approach ensures a comprehensive assessment of airfoil efficiency, especially in low-speed wind turbine scenarios. Focused on a consistent low wind speed (Re = 10^5) and a fixed 15-degree AoA, our research aims to enhance the airfoil's aerodynamic characteristics. Following optimization, the morphed profile undergoes rigorous evaluation using XFLR5 software, exploring behaviour across various angles and Re. Through extensive investigations, our study delivers an optimized morphed profile for the subject airfoil, promising heightened efficiency, and reliability in low-speed conditions.

2 Methodology

The methodology consists of three main steps, PARSEC parametrization of NACA2414 airfoil, GA optimization for maximum lift to drag ratio and validation of optimized profile using XFLR5 solver. Each step as shown in **Fig. 1**, will be elaborated further in this section.

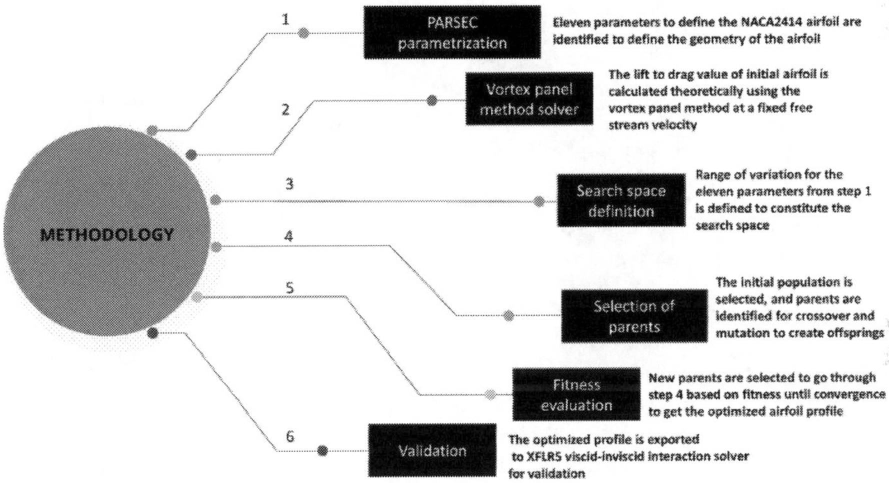

Fig. 1. Research methodology

2.1 PARSEC parametrization

PARSEC is a widely recognized method for effectively representing airfoil shapes. It operates by combining basic functions linearly and relies on a set of eleven distinct geometric properties of the airfoil as its foundation. These properties, which also serve as design variables within the PARSEC framework, are essential for describing the airfoil's shape. By utilizing these design variables, PARSEC allows for the creation of a system of linear equations, which accurately defines the airfoil's geometry. This adaptable parameterization approach plays a crucial role in optimizing airfoil designs and improving their aerodynamic performance as seen from the literature. The parameters for the specific airfoil, NACA2414, under consideration are listed in **Table 1** and can be better understood when referenced alongside **Fig. 2.**

Table 1. PARSEC parameters and specified range for the NACA2414 airfoil

	Initial parameters	Range
r_{le}	0.0216	0.0011
X_{up}	0.3240	0.0162
Z_{up}	0.0888	0.0044
Z_{xxup}	-0.6363	-0.0318
X_{lo}	0.2365	0.0118
Z_{lo}	-0.0519	-0.0026
Z_{xxlo}	0.3141	0.0157
Z_{TE}	0	0
ΔZ_{TE}	0	0
α_{TE}	0.0332	-0.950
β_{TE}	0.3253	9.324

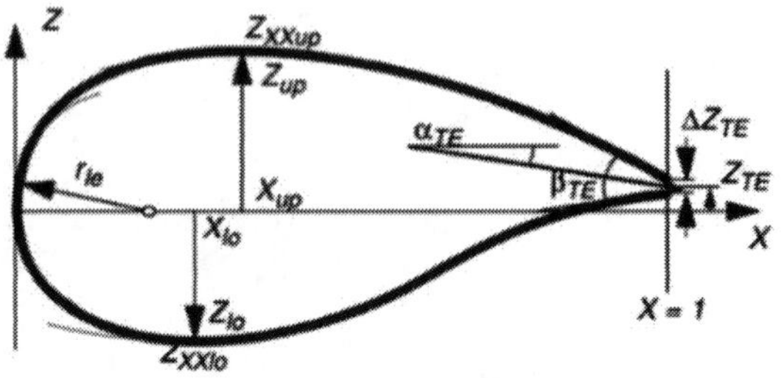

Fig. 2. PARSEC parameters for an airfoil [5]

The notations shown in **Fig. 2** are explained below:

r_{le} : leading edge radius
X_{up} : X coordinate for upper surface (max. position)
X_{lo} : X coordinate for lower surface (max. position)
Z_{up} : Z coordinate for upper surface (max. position)
Z_{lo} : Z coordinate for lower surface (max. position)
Z_{xxup} : curvature in the position of upper surface (max. thickness)
Z_{xxlo} : curvature in the position of upper surface (max. thickness)
α_{TE} : inclination angle of mean camber line in the trailing edge (degree)
β_{TE} : separation angle at trailing edge (degree)
Z_{TE} : Z coordinate of trailing edge
ΔZ_{TE} : Separation between upper and lower surface in trailing edge

The selection of the range for airfoil parameters is carefully tailored with a particular focus on making significant adjustments to the airfoil's trailing edge. While allowing a 5% variation range for all other parameters, α_{TE} and β_{TE} are set with a broader 50% range. Notably, the trailing edge thickness is maintained at zero during the optimization process. These variables can be adjusted to produce various transformed airfoil profiles. However, this study places a primary emphasis on the critical alterations needed at the trailing edge. This is pivotal, particularly in low-speed conditions, where it plays a crucial role in managing flow separation and wake characteristics.

2.2 Genetic Algorithm optimization

GAs provide an innovative, non-gradient optimization approach, inspired by the theory of evolution. In this method, chromosomes are ranked based on their fitness vector, with higher-ranking ones progressing to the next generation. This evolution process involves creating new sets of artificial creatures using components from the fittest of the previous generation, introducing occasional new elements. GAs efficiently utilizes historical information to speculate on improved search points, incorporating operators like combination, crossover, or mutation to manipulate the newly selected chromosomes. The process iterates through generations until a suitable level of convergence is achieved or a specified number of generations is completed. GAs excels in non-smooth design spaces with multiple local extrema and are particularly useful for multi-objective design optimization. The original GA code used in the study can be found in [16] which was modified for optimizing the lift to drag ratio of NACA2414 airfoil. The aerodynamic lift and drag forces were calculated using the vortex panel method as described in the next section and the ratio between lift to drag was chosen to be the objective function for the optimization procedure.

2.2.1 Vortex panel method

The vortex panel method, employed in this study for design optimization utilizes discretization of the airfoil surface into straight line segments or panels. The method assumes constant source strength over each panel with varying values and constant vortex strength across panels. The approach neglects air compressibility and viscosity while ensuring the net vorticity at the trailing edge maintains smooth flow. The numbering system starts at the lower surface trailing edge and proceeds around the leading surface and behind to the upper surface trailing edge. Flow tangency boundary condition is imposed at the midpoint of each panel, facilitating the computation of pressure coefficients for optimization. Its primary advantage lies in its applicability to simplify fluid motion analyses for engineers and scientists. By focusing on ideal flows, the method provides a foundational understanding of fluid behaviour without the complexities introduced by compressibility and viscosity. Its simplicity and effectiveness in handling these simplified scenarios make it an excellent starting point for comprehending fluid dynamics principles.

2.2.2 Search space definition

In the context of the GA employed in this study, the search space is meticulously defined by the range of values assigned to the PARSEC parameters, each adhering to a specified accuracy. The optimization parameters are characterized by the required precision that enables the expression of the domain length for each optimization parameter. The careful consideration of these parameters and their defined search space lays the foundation for an effective and targeted optimization process. Achieving the necessary precision for optimization parameters involves defining the permissible range or boundaries for each parameter, considering factors like system limitations and engineering constraints. This definition includes specifying the step size or granularity within the range to control optimization search resolution. Striking a balance between granularity and computational efficiency is crucial, as too fine a granularity may lead to excessive resource use, while too coarse a granularity may overlook optimal solutions. Through careful consideration and definition of domain lengths, the optimization process navigates the parameter space effectively, ensuring the exploration of solutions that meet the required precision criteria.

2.2.3 Selection of parents

In the algorithm employed, the selection of individuals from the initial population is a crucial step, and the tournament selection technique is utilized. This approach involves organizing individuals into pairs, where candidates compete against each other, and the one with superior fitness is chosen to advance to the next generation. Notably, the selection process maintains variability by not excluding the worst individuals. The subsequent generation is then generated through crossover, involving the combination of desirable traits from two parents chosen for mating. Additionally, mutation plays a role in refining the mating process, introducing randomness by flipping bits based on a mutation probability. The careful orchestration of these steps contributes to the efficiency and diversity of the optimization.

2.2.4 Fitness evaluation

Fitness evaluation is a pivotal stage in the optimization process, involving the assessment of the objective function for each set of optimization parameters. The fitness of the newly generated offspring determines their eligibility as potential parents for further mating. This iterative process persists until the optimization converges, signifying the attainment of an optimal solution. The efficacy of the GA hinges on the accurate evaluation of fitness, ensuring the selection of individuals with favourable traits for subsequent generations.

2.3 Validation

The aerodynamic shape optimization process aims to enhance the lift to drag ratio of the chosen airfoil. Given the applicability of the panel method to low-speed flows, a flow constraint is imposed to maintain validity throughout optimization. Constraints are incorporated, such as the thickness of the airfoil is set to a minimum value of 0.1, to prevent unrealistic shapes. PARSEC parameters describe the airfoil shape, with GA optimizing within specified bounds. Each generation yields the best set of parameters, generating corresponding airfoil profiles using PARSEC parameterization. The airfoil undergoes testing at a 15-degree angle of attack in subsonic, incompressible flow. The pressure distribution informs the coefficient of lift to drag ratio, compared to the original. The design conditions and optimization objective are detailed in **Table 2**. The transcendence percentage, crossover percentage, and mutation percentage are applied to the population size to determine the number of individuals involved in each operation.

The lift-to-drag ratio is optimized at a 15-degree angle of attack, with a free stream velocity of 1.4607 m/s. The optimized profile, generated in MATLAB, is exported to XFLR5 for evaluation across two different low Re and a range of AoAs from -10 to 15 degrees.

Table 2. Design conditions and optimization parameters for GA pipeline

Number of generations to mate	20
Free stream velocity	1.4607 m/s (corresponding to Re = 10^5)
Population size	48
Transcendence percentage	0.05
Cross over percentage	0.75
Mutation percentage	0.2
Angle of attack	15-degree
Optimizing function	Lift to drag ratio

3 Results and discussion

In the upcoming section, we explore the NACA 2414 airfoil's performance under varying conditions, with a focus on lift and drag coefficients, along with the lift-to-drag ratio. Analysing two Re, 10^5 and 5×10^5, representing distinct aerodynamic scenarios, aims to reveal the airfoil's response to varying Re and its behaviour in different flow conditions. The findings will offer valuable insights into the airfoil's aerodynamic characteristics, especially in low-speed wind turbine operations. The selected range of AoAs is crucial for wind turbine airfoils, governing their aerodynamic performance and energy conversion efficiency. Wind turbine blades, rotating through a spectrum of AoAs, necessitate understanding the airfoil's behaviour within this range. The balance between maximizing lift for power generation and minimizing drag is crucial, with the AoA range considered from -10 to 15 degrees in this study. Significantly, at a 15-degree AoA, the airfoil approaches its stall angle, highlighting the importance of this specific range for optimizing aerodynamic performance in low-speed wind turbine operations. The original airfoil and the optimum profile airfoil are represented in **Fig. 3.**

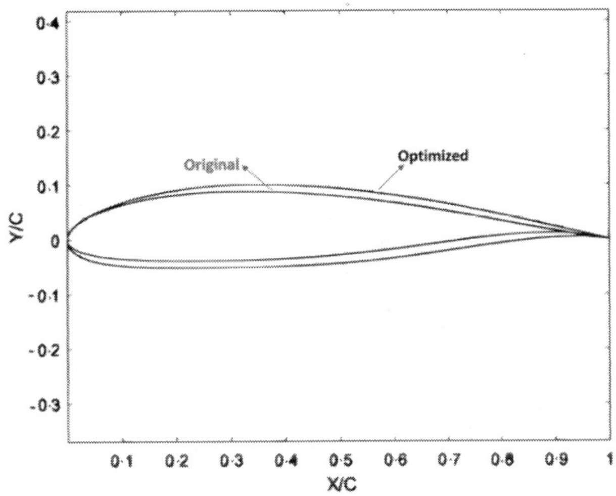

Fig. 3. Comparison of original and optimized NACA2414 airfoils

The findings of this study revealed substantial improvements in the lift coefficients of the NACA2414 airfoil as shown in **Fig. 4(a)**, after undergoing optimization with GA. When we look at different AoAs, it is observed that the optimized airfoil consistently outperforms the original one across the entire range. At a 15-degree AoA, which is of prime importance being close to the stall angle of the particular airfoil and control point for the optimization process, the lift coefficient of the optimized airfoil reaches 1.5275, while the original airfoil achieves only 1.2407 at a Re of 10^5. This implied that the optimized airfoil produced substantially greater lift, a critical factor for its suitability in low-speed wind environments for higher power generation. The results presented for Re 5×10^5 showed a similar trend. At varying AoAs, the modified airfoil consistently outperformed the original airfoil as can be seen from **Fig. 4(b).**

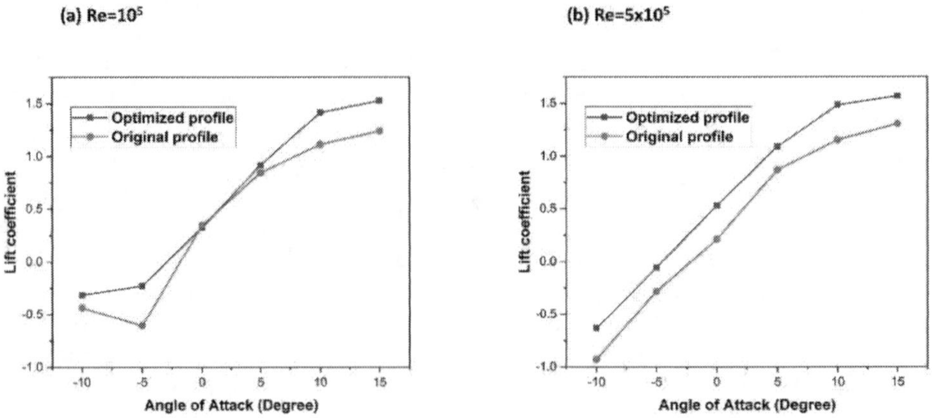

Fig. 4. Lift coefficients comparison of original and optimized airfoils at Reynolds numbers 10^5 and $5{\times}10^5$

The drag coefficient results for Re of 10^5 display a noteworthy pattern as displayed in **Fig. 5(a).** When considering different AoAs, the modified airfoil exhibits similar variation until stall approaches. The drag coefficient is higher at AoA of 15-degree at 0.0764 compared to the original profile drag at 0.0646, while this indicates a modest increase, it's essential to recognize that the primary focus of the optimization process was on enhancing the lift to drag ratio. The relatively small differences in drag coefficients are outweighed by the significant improvements in lift. The drag coefficient variation at Re of $5{\times}10^5$ as depicted in **Fig. 5(b).** exhibited similar pattern but at the AoA of 15-degree, this increase was substantially higher.

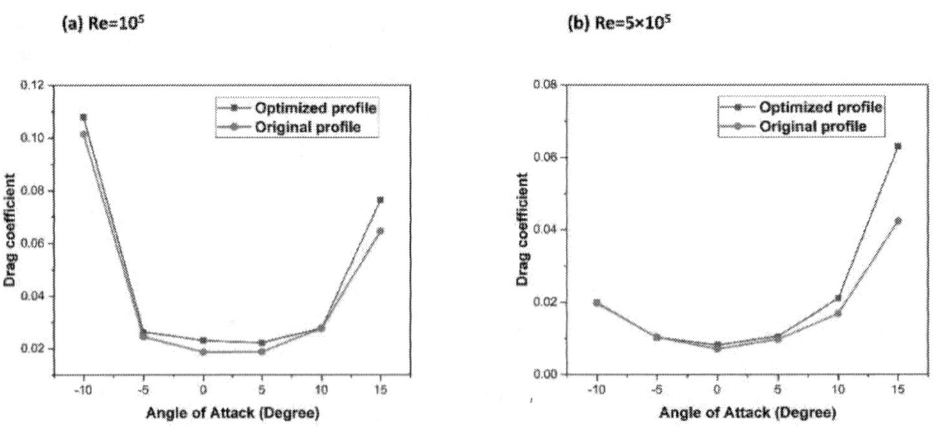

Fig. 5. Drag coefficients comparison of original and optimized airfoils at Reynolds numbers 10^5 and $5{\times}10^5$

The lift-to-drag ratio, a critical indicator of aerodynamic efficiency, showcases notable improvements in the optimized airfoil compared to the original NACA2414 airfoil at specific AoAs and is plotted in **Fig. 6(a)** for a Re of 10^5. At 10-degree AoA, the lift-to-drag ratio for the modified airfoil is significantly higher, with a value of 50.78, compared to 40.08 for the original airfoil. This substantial increase underscores the effectiveness of the GA optimization in enhancing aerodynamic efficiency. Similarly, at 15-degree AoA, the lift-to-drag ratio for the modified airfoil is 19.99, while the original airfoil has a ratio of 19.22. The enhancements observed at higher AoAs are particularly remarkable as they signify the optimized airfoil's capability to sustain its performance advantage even in more demanding conditions. The alterations in the lift-to-drag ratio at Re of $5{\times}10^5$ are depicted in **Fig. 6(b).**

At an AoA of 15-degree, the modified airfoil demonstrated a lift-to-drag ratio of 24.95, while the original airfoil exhibits a slightly higher value of 30.88. This finding indicates a minor decrease in the lift-to-drag ratio for the modified airfoil at this specific AoA. The optimized airfoil has shown improvements at other AoAs regardless of this. It's essential to consider the overall trade-off between different AoAs and lift coefficients when evaluating the airfoil's performance.

In summary, optimizing the NACA2414 airfoil using PARSEC parameters and GA significantly improved lift coefficients, particularly at the critical 15-degree AoA. This enhancement, observed across different Re, highlights the airfoil's suitability for efficient power generation in low-speed wind environments. Notably, the optimization prioritized the lift-to-drag ratio, leading to substantial improvements. At 10-degree AoA and Re 10^5, the modified airfoil showed a significantly higher lift-to-drag ratio compared to the original, emphasizing effectiveness of the method in enhancing aerodynamic efficiency. Despite a minor decrease in the lift-to-drag ratio at 15-degree AoA and Re 5×10^5, the overall findings underscored the optimized airfoil's ability to maintain performance advantages, particularly in challenging conditions. These results highlight the success of the optimization process in enhancing the airfoil's aerodynamic efficiency, with specific attention to the lift-to-drag ratio.

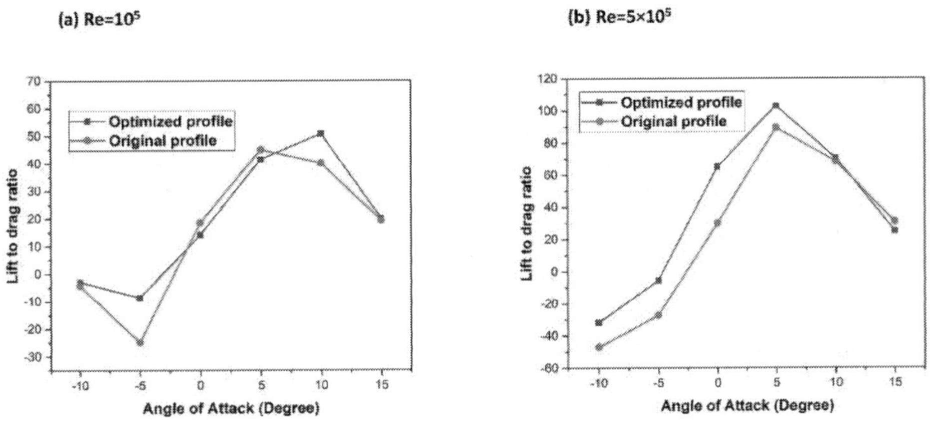

Fig. 6. Lift to drag ratio of original and optimized airfoils at Reynolds numbers 10^5 and 5×10^5

4 Conclusion

In the pursuit of improving the aerodynamic performance of the NACA2414 airfoil, this study employed PARSEC parametrization and Genetic Algorithm optimization, specifically targeting an enhanced lift-to-drag ratio across angles of attack (-10 to 15 degrees). Results revealed a substantial increase in lift coefficients after Genetic Algorithm optimization, emphasizing the effectiveness of the approach. Notably, the optimization process significantly improved the lift-to-drag ratio, with the modified airfoil outperforming the original, particularly at 10 and 15-degree angles of attack. At 10-degree AoA, the modified airfoil achieved a notably higher lift-to-drag ratio of 50.78, compared to 40.08 for the original airfoil at Reynolds number 10^5. The improvements in efficiency extended to 15-degree AoA, where the lift-to-drag ratio for the optimized airfoil reached 19.99, while the original airfoil had a ratio of 19.22 at Reynolds number of 10^5. The success of the optimization was tied to Reynolds number and parameter selection. These findings highlight the potential for Genetic Algorithm techniques to enhance wind turbine airfoil efficiency, particularly in low wind speeds. The study underscores the value of incorporating advanced methods, such as

Computational Fluid Dynamics simulations and machine learning algorithms, in airfoil optimization for continued advancements in renewable energy technologies.

Abbreviations

AoA : Angle of attack
GA : Genetic Algorithm
Re : Reynolds number

Declaration of Competing Interest

The authors declare that they have no known competing financial interests or personal relationships that could have appeared to influence the work reported in this paper.

Data availability

The data used for the research will be made available upon request.

Acknowledgement

This work was supported by Taylor's University through its Taylor's Research Scholarship Programme.

References

1. Suprayitno, J. C. Yu, Aminnudin, and R. Wulandari, *Airfoil aerodynamics optimization under uncertain operating conditions*, in The 2nd International Conference on Vocational Education of Mechanical and Automotive Technology, 12 October 2019, Yogyakarta, Indonesia (2019).
2. W. Chen, K. Chiu, and M. Fuge, AIAA Journal **58(11)**, 4723-4735 (2020).
3. M. S. Khurana, H. Winarto, and A. K. Sinha, *Airfoil Optimisation by Swarm Algorithm with Mutation and Artificial Neural Networks,* in 47th AIAA Aerospace Sciences Meeting Including The New Horizons Forum and Aerospace Exposition, 5 - 8 January 2009, Orlando, Florida (2009).
4. G. Tortora, A. Concilio, and R. Pecora, Designs **7**, 28 (2023).
5. R. Mukesh, K. Lingadurai, and U. Selvakumar, Journal of King Saud University - Engineering Sciences **26(2)**, 191-197 (2014).
6. N. P. Salunke, J. Ahamad R. A., and S. A. Channiwala, American Journal of Mechanical Engineering **2(4)**, 99-102 (2014).
7. A. Krishnan, A. S. M. Al-Obaidi, and L. C. Hao, Sustainable Energy Technologies and Assessments **60**, 103511 (2023).
8. H. C. Ayazümütlü and Z. Kiral, Aviation **26**, 32-40 (2022).
9. A. Michelotti, A. Cavini, R. Giacopino, F. Misino, and L. Piottoli, *Implementation of genetic algorithms for supersonic airfoil optimization*, in International Conference on Innovation in Aviation & Space to the Satisfaction of the European Citizens (11th EASN 2021), 1 – 3 September 2021, Salerno (2021).
10. D. Viet Dung, V. Dung Duong, and V. Anh Duong, JST: Smart Systems and Devices **33(3)**, 49-56 (2023).
11. H. Kamliya Jawahar, Q. Ai, and M. Azarpeyvand, Renew Energy **127**, 355-367 (2018).
12. M. Baghdadi, S. Elkoush, B. Akle, and M. Elkhoury, Renew Energy **154**, 239-251 (2020).
13. R. A. Leonczuk Minetto and M. Paraschivoiu, Energy **202**, 117705 (2020).

14. H. Jaffar, L. Al-Sadawi, A. Khudhair, and T. Biedermann, International Journal of Thermofluids **17**, 100267 (2023).
15. Y. C. Li, J. J. Wang, and J. Hua, Aerosp Sci Technol **11**, 91-99 (2007).
16. Z. Saad, Ziad Saad (2023), Optimization of an airfoil shape using Genetic algorithm(GA) (https://www.mathworks.com/matlabcentral/fileexchange/69095-optimization-of-an-airfoil-shape-using-genetic-algorithm-ga), MATLAB Central File Exchange. Retrieved September 23, 2023.

Efficiency of Heating and Evaporation of Graphene Nanofluid under Solar Radiation

Quoc Thinh Tran [1*], *Alexander* Dmitriev[1], *Inna* Mikhailova[1]

[1]National Research University "Moscow Power Engineering Institute", 111250 Moscow, Russia

Abstract. The conversion of solar radiation into steam is a crucial focus in today's green energy, ecology, and clean water production. This study presents the first investigation of the heating and evaporation process of a rotating graphene nanofluid under the influence of solar simulator radiation. The study examined the influence of various factors on the heating and evaporation process of the graphene nanofluid, including the direction of irradiation, graphene concentration and rotation speed. It was demonstrated that the evaporation rate strongly depends on the graphene concentration and the irradiation method of the samples. The heating characteristics of graphene also depend on the irradiation method. It was shown that graphene heated to a higher temperature when in direct contact with radiation, while graphene within the bulk volume was heated less effectively than the base liquid. Moreover, the application of rotating graphene nanofluid in this research was found to enhance thermal efficiency by 2.5% compared to distilled water, with a graphene volume concentration of 0.1%. Consequently, various effects of the rotating graphene nanofluid volume on hydrodynamic, heat, and mass transfer parameters were identified, which hold significance for both fundamental and applied tasks in energy, chemical technology, and ecology.

1 Introduction

Currently, special attention is being given to the research of nanofluids in the context of solar thermal energy conversion, aiming to convert solar energy into steam for subsequent electricity generation or production of clean, desalinated water [1-3]. Over the past 10 years, the number of studies on this topic has been continuously increasing [4]. Scientists have attempted to enhance the efficiency of energy systems by utilizing nanofluids as working fluids [5-7]. However, there are several challenges that need to be addressed for the successful application of such a heat transfer medium. For instance, in solar thermal energy systems, in order to achieve efficient conversion of solar radiation into steam, it is necessary to develop absorbers (nanoparticles) that exhibit broadband light absorption. This characteristic, in turn, depends on the properties of the nanoparticles, including material, shape, concentration, sizes, and their interaction with the base fluid [8-10]. Additionally, the efficiency of the solar-driven steam generation process also depends on external factors such as the direction of radiation, optical thickness of the medium, and device design [11].

* Corresponding author: tranqth.96@gmail.com

This study is dedicated to investigating the efficiency of heating and evaporation processes of rotating graphene nanofluid under the influence of solar radiation, which can be beneficial for water purification systems. The influence of various factors on the heating and evaporation process is examined, including graphene concentration, rotation speed, and irradiation direction. It is demonstrated that higher heating and steam generation efficiency can be achieved by utilizing a rotating volume of graphene nanofluid. However, it is observed that the graphene nanofluid is unstable during rotation and heating, which limits the potential for enhancing the heating and evaporation process efficiency. The obtained results can be utilized in the development of water purification systems using a rotating volume of working fluid under solar radiation and serve as a basis for exploring new possibilities and creating novel graphene-based designs to enhance the efficiency of solar steam generation process.

2 Experimental setup and methods

The overall view of the experimental setup is depicted in Fig. 1. In this study, the rotation of graphene nanofluid is achieved using a magnetic stirrer MSH-20D Daihan with an 8 mm diameter and 25 mm length magnetic stir bar. Graphene nanoflakes were obtained through exfoliation (splitting) of high-purity graphite (purity of 99.2%) in distilled water. The graphene nanoflakes were characterized using Raman spectroscopy and scanning electron microscopy. The number of graphene layers ranges from 1 to 10, with an approximate 40% average of graphene nanoflakes having fewer than 5 layers. The lateral dimensions of graphene flakes range from 30 to 90 μm (Fig. 2B.4). Raman spectroscopy has revealed that the graphene flakes are non-oxidized, with the presence of graphene oxide not exceeding 5%. The experiments were conducted on graphene nanofluid contained in a cylindrical vessel with an internal diameter of 44 mm, wall thickness of 1.5 mm, and height of 100 mm. The initial mass of the samples is 100 g. The heating of the samples was carried out using a half-quartz infrared heating element ICH-402, which serves as a solar simulator. This simulator has a specific surface radiation energy of 1.91 W/cm^2 and an effective surface area of 78.4 cm^2. It generates radiation in the range of 1.85 to 8.9 μm. The measurement of the temperature field on the surfaces of the samples (Fig. 1A) was performed using the Seek Thermal Compact Pro thermal camera, which is capable of measuring temperatures in the range of -40 to 330°C. It has a sensitivity of less than 70 mK and a resolution of 320 × 240 pixels. The temperature inside the volume of the samples was measured using the SS200 thermocouple (Fig. 1B) at different graphene concentrations and different magnetic stirrer rotation speeds (ranging from 0 to 900 rpm). The amount of evaporated water was measured using electronic microbalances with an accuracy of 0.01 g.

The preparation of the graphene nanofluid was conducted using a two-step method involving ultrasonic treatment for 40 minutes. This process aimed to disaggregate large clusters of graphene flakes and achieve their uniform dispersion in the base liquid (distilled water) without the addition of surfactants. To obtain reliable results, each experiment with every graphene concentration was conducted sequentially with high quality and repeated five times. The maximum relative error in calculating the evaporation rate, in the case of samples irradiated from the side, is 9.1%, while it is 16.5% when irradiated from the top. In order to minimize the influence of the surrounding environment (such as wind, temperature, and humidity changes) on the research outcomes, the experiments were conducted in a sealed room under the following conditions: air temperature in the room was maintained at 28 ± 1°C, the initial temperature of the samples was 28 ± 0.5°C, and the relative humidity in the room was 40 ± 5%.

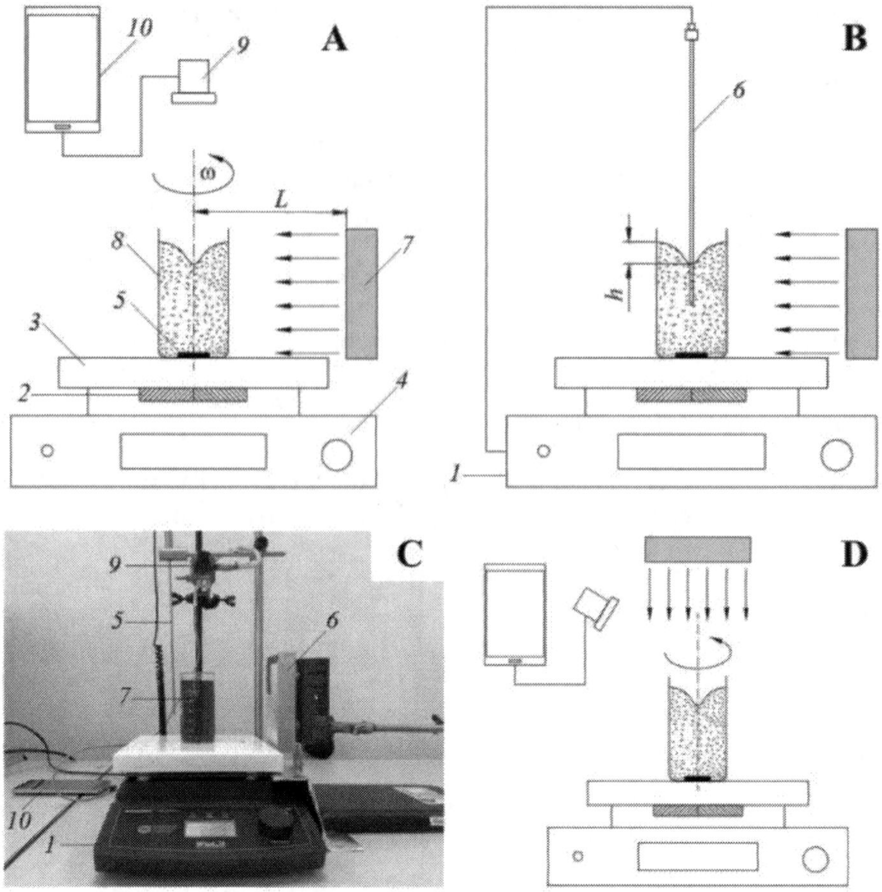

Fig. 1. Diagram (A, B) and photograph (C) of the experimental setup used for measuring the surface temperature and temperature within the volume of graphene nanofluid under side irradiation. Diagram for measuring the surface temperature under top irradiation (D): 1 – magnetic stirrer MSH-20D Daihan; 2 – stirrer magnet; 3 – ceramic platform for placing the sample; 4 – stirrer control panel; 5 – magnetic stir bar; 6 – SS200 thermocouple; 7 – half-quartz infrared heating element ICH-402; 8 – cylindrical vessel containing the graphene nanofluid under investigation; 9 – Seek Thermal Compact Pro thermal camera connected to a phone 10.

3 Results and discussion

3.1 The influence of graphene concentration, irradiation direction, and rotation speed on the evaporation process

The efficiency of the evaporation process for the studied samples was evaluated through the average evaporation rate, denoted as j (mg/s). The influence of the following factors on the evaporation process was considered: volumetric concentration of graphene nanoparticles, irradiation direction, and rotation speed of the nanofluid volume. The investigation of the influence of volumetric concentration of graphene and irradiation direction was conducted at a magnetic stir bar rotation speed of 600 rpm. The obtained results regarding the influence of the studied factors on the average evaporation rate of the samples are presented in Fig. 2A.

 https://doi.org/10.1051/e3sconf/202451403005

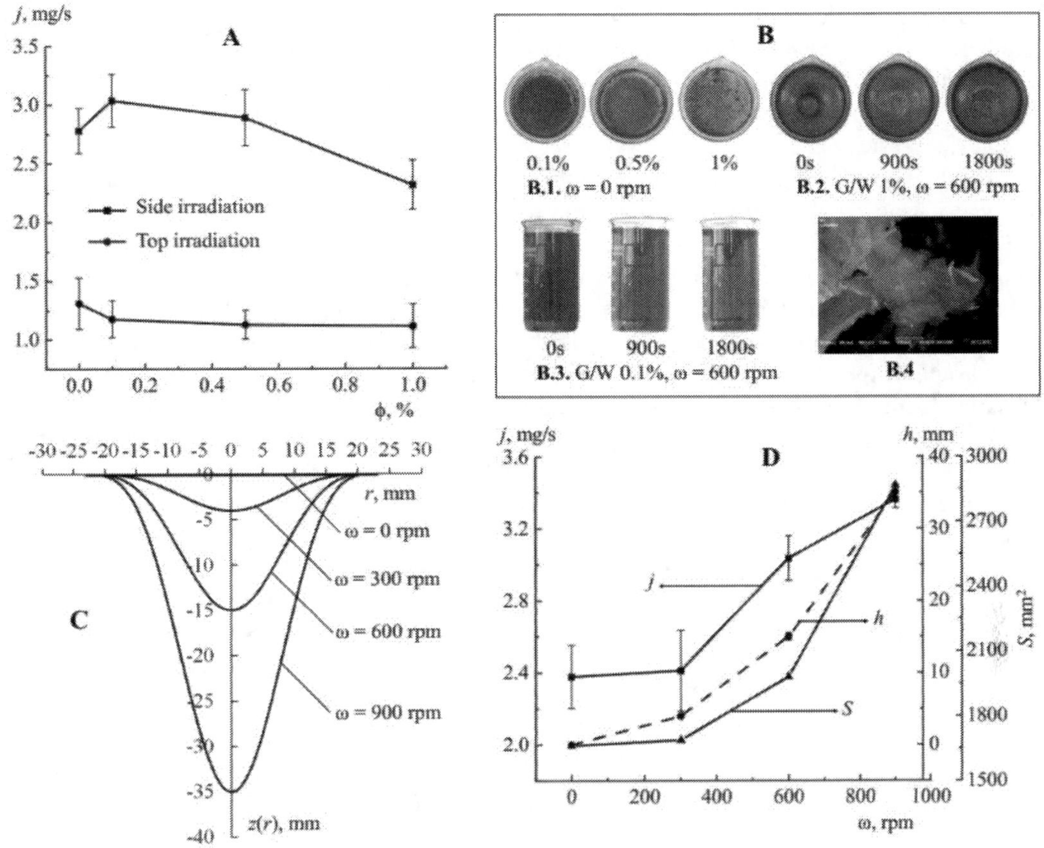

Fig. 2. The influence of graphene concentration and irradiation direction on the average evaporation rate (A). Morphology of the graphene nanofluid surface before rotation and irradiation (B.1) and during the rotation and irradiation process (B.2); coagulation and formation of aggregates from graphene nanoflakes in the lower part of the funnel (B.3) over time at a volume concentration of 0.1% and a rotation speed of 600 rpm; SEM image of graphene nanoflakes (scale bar: 10 μm) (B.4). Profile of the interfacial surface of the rotating liquid at different rotation speeds (C). Influence of the magnetic stir bar rotation frequency on the interfacial surface area *S* and evaporation rate *j* (D).

It can be observed that in the case of side irradiation of the samples (where radiation is transmitted through the glass wall into the nanofluid), the evaporation rate increases with the increase in graphene concentration, reaching its maximum value at a volumetric concentration of 0.1%, and then decreases as the concentration is further increased within the range of 0.5 to 1.0%. It should be noted that compared to distilled water, an increase in the evaporation rate is observed only for graphene nanofluid with volumetric concentrations of 0.1% and 0.5%, with values of 9% and 4%, respectively. At a concentration of 1.0%, the graphene nanofluid exhibited an evaporation rate that was 16.5% lower than that of distilled water. In the case of top irradiation (where radiation directly falls on the surface of the graphene nanofluid), the evaporation rate for all samples was lower than that of distilled water. This behavior can be attributed to the characteristics of the graphene film formed at the interface between the graphene nanofluid and air phases (Fig. 2B). It was observed that with an increase in concentration, the quantity of graphene nanoflakes in the film increased, leading to a significant reduction in the free interface area between the water and air phases (Fig. 2B.1). Additionally, during the rotation and heating process, strong coagulation of graphene nanoflakes occurred, resulting in the formation of large aggregates. During rotation, the formed aggregates gather at the lower region of the interfacial surface between the graphene nanofluid and air phases, and over time, their quantity increased. This can be clearly seen in the example of the sample with a concentration of 0.1% (Fig. 2B.3). At higher graphene concentrations (0.5% and 1.0%), the coagulation process led to a significant change in shape and a reduction in the interfacial area between the graphene nanofluid and air phases (Fig. 2B.2). As a result, these factors reduced the mass flow of evaporating water vapor from the interfacial surface.

The shape of the interfacial surface between the graphene nanofluid and air phases and the evaporation rate strongly depend on the rotation speed of the magnetic stir bar ω (Fig. 2C). Figure 2D presents experimental data on the dependence of the "funnel" height *h* and evaporation rate on ω, along with the theoretical relationship between the interfacial area *S* and ω. It has been found that as ω increases from 0 to 300 rpm, the evaporation rate increases insignificantly. The increase in the evaporation rate in this case is only 1.4%. However, at higher rotation speeds, the evaporation rate sharply increases. At ω equal to 600 and 900 rpm, the evaporation rate increases by 27.7% and 41.3% respectively. This is attributed to the fact that as ω increases to 600 and 900 rpm, the height of the "funnel" *h* increases significantly, resulting in a substantial increase in the interfacial area between the graphene nanofluid and air phases and, consequently, an increase in the mass flow rate of evaporating water. Moreover, at higher rotation speeds, the heat transfer process within the sample volume improves, and water molecules possess greater kinetic energy, which also contributes to the enhancement of the evaporation process. Below are the formulas that have been derived from solving the Navier-Stokes momentum conservation equations to calculate the profile and surface area of a rotating liquid's interfacial surface, as a function of the funnel height *h* (or rotation speed ω).

$$z(r,t) = h\left\{\frac{1-\left[J_0^2\left(\frac{\mu_1}{R}r\right)+J_1^2\left(\frac{\mu_1}{R}r\right)\right]e^{-2\nu\frac{\mu_1^2}{R^2}t}}{1-J_0^2(\mu_1)} - 1\right\}, \quad (1)$$

$$S = \int_0^{2\pi}\int_0^{R}\sqrt{1+\left(\frac{\partial z}{\partial r}\right)^2+\left(\frac{\partial z}{\partial \varphi}\right)^2}\, r dr d\varphi, \quad (2)$$

where $J_0(x)$ and $J_1(x)$ are the Bessel functions of the first kind of zeroth and first order, respectively; $\mu_1 = 3.8317$ is the first root of the equation $J_1(\mu) = 0$; *r* is distance from the axis of symmetry of the cylindrical vessel to the point under consideration; *R* is the inner radius of the vessel; ν is the kinematic viscosity of the fluid; *h* is the height of the funnel; *t* is the time.

3.2 The effect of heating in a rotating graphene nanofluid

The heating process of the graphene nanofluid were investigated by varying the volume concentration of graphene nanoflakes, irradiation direction, and magnetic stir bar rotation speed. Fig. 3 shows the temperature field of the interfacial surface between the fluid and air phases after 30 minutes of side and top irradiation. It can be observed that in the case of side irradiation the temperature fields of the water surface and the graphene nanofluid surface differ significantly from each other. In particular, the temperature values of the graphene nanofluid surface were considerably lower than those of the water surface. Over a duration of 30 minutes, the interfacial surface between water and air was heated to approximately 95-100°C (Fig. 3A.1), whereas the interfacial surface between graphene nanofluid and air attained temperatures ranging from 80-87°C. Furthermore, the water surface is heated uniformly, while the temperature field of the interfacial surface between graphene nanofluid and air becomes non-uniform due to the non-homogeneity of the graphene film with increasing graphene concentration. It is worth noting that the lower part of the interfacial surface, where a large amount of coagulated graphene accumulates, has the lowest temperature (Fig. 3A.2, A.3, and A.4). Therefore, heat transfer between coagulated graphene and water is not efficient. However, there was a notable shift in the situation when the sample was subjected to top irradiation. In this case, the interfacial surface between fluid and air directly interacts with the incident radiation. The findings indicate that the graphene film on the interfacial surface achieved elevated temperatures of approximately 130-150°C, surpassing the temperature of both the water surface and the vessel wall to a significant degree (Fig. 3B.1-4).

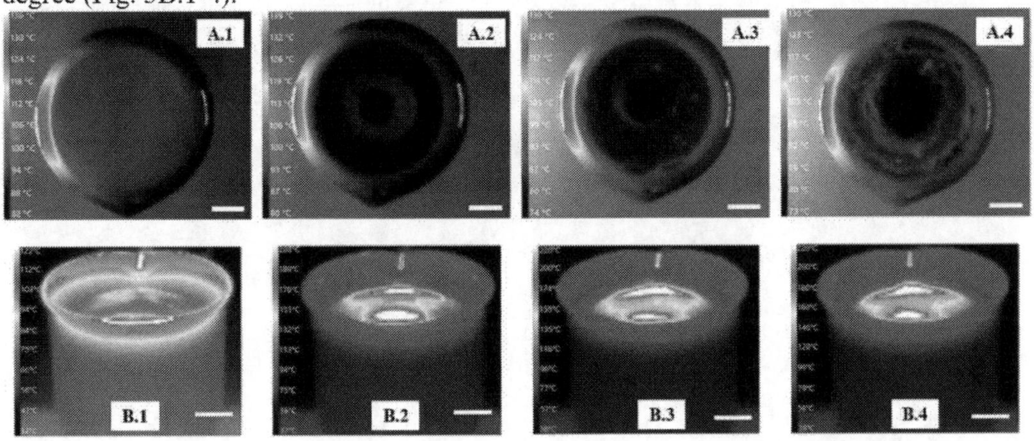

Fig. 3. Temperature field of the interfacial surface between fluid and air after 30 minutes of side irradiation: distilled water (A.1); G/W 0.1% (A.2); G/W 0.5% (A.3); G/W 1.0% (A.4). Temperature field of the sample surfaces after 30 minutes of top irradiation: distilled water (B.1); G/W 0.1% (B.2); G/W 0.5% (B.3); G/W 1.0% (B.4) (scale bar: 1 cm).

3.3 The efficiency of converting solar radiation into thermal energy

The efficiency of converting solar radiation into useful thermal energy, including the process of vaporization (η_{evap}) and heating (η_{heat}) of the working fluid volume, is determined by the following equation:

$$\eta_T = \eta_{heat} + \eta_{evap} = \frac{c_p(T_m^b - T_0) \cdot (m_0 - \Delta m) + L\Delta m}{F \cdot q_{solar} \cdot \Delta t}, \quad (3)$$

where T_0, m_0 represents the initial temperature and mass of the liquid volume; Δm is the amount of liquid evaporated during the time interval Δt; c_p is the specific heat capacity; L is the latent heat of vaporization, which is equal to 2256 kJ/(kg) at atmospheric pressure for water; F is the irradiated surface area; T_m^b is the average temperature of the liquid volume after the time interval Δt of irradiation; q_{solar} is the incident radiation flux density on the surface of the sample, which can be estimated using the angular emissivity coefficient φ_{12} according to equation (4):

$$q_{solar} = \varphi_{12} \cdot q_0, \quad (4)$$

here q_0 represents the specific radiation flux density of the source, measured in W/m². For the investigated system, which consists of an emitting surface of rectangular shape 1 (Fig. 4) and a directly positioned circular cylinder 2 symmetrically relative to surface 1, with a base radius of R and a height equal to the length of side a of the rectangular surface. In this case, the angular emissivity coefficient φ_{12} is determined by formula (5):

$$\varphi_{12} = 2\frac{H}{B}\int_{0}^{B/2} F(Y)\mathrm{d}Y. \quad (5)$$

The function *F*(*Y*) and the parameters *B*, *H* in equation (5) are determined by the following relationships

$$F(Y) = \frac{1}{H^2+Y^2}\left\{1-\frac{1}{\pi}\arccos\left(\frac{Z-Y^2}{X+Y^2}\right)-\frac{2}{\pi A}\left[\sqrt{X+Y^2+4A^2}\cdot\right.\right.$$
$$\left.\left.\cdot\arccos\left(\frac{Z-Y^2}{(X+Y^2)\sqrt{H^2+Y^2}}\right)+\frac{1}{\pi}(Z-Y^2)\arcsin\left(\frac{1}{\sqrt{H^2+Y^2}}\right)-\frac{X+Y^2}{2}\right]\right\},$$

where

$$A=\frac{a}{R}; B=\frac{b}{R}; H=\frac{h}{R}; X=A^2+H^2-1; Z=A^2-H^2+1.$$

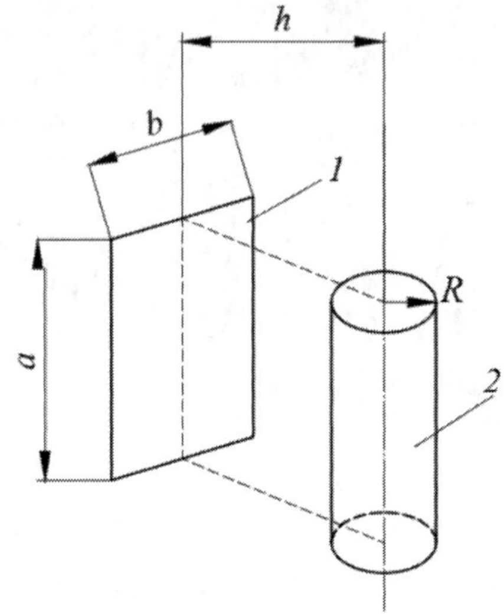

Fig. 4. The diagram for calculating the angular emissivity coefficient of a system consisting of a rectangular surface 1 and a cylindrical vessel 2.

In this study, a solar simulator with dimensions of $a\times b$, equal to 100 mm × 60 mm, and a cylindrical vessel with a height of 100 mm and a radius of 23 mm were used. According to equation (5), it was determined that the angular emissivity coefficient φ_{12} is equal to 0.197. Considering that the specific energy of the surface radiation of the simulator q_0 is 19100 W/m^2, the incident radiation flux density on the surface of the cylindrical vessel q_{solar} is equal to 3763 W/m^2. The data on the temperature difference and the amount of evaporated water after 30 minutes of side irradiation of rotating graphene nanofluids at a rotation speed of 600 rpm were experimentally obtained and presented in Tab. 1. The specific heat capacity of graphene nanofluids is estimated using equation (6). Experimental data in the study [12] have shown that in the temperature range up to 200°C, the specific heat capacity of graphene and graphite is approximately the same and is around 787 J/(kg·K). Additionally, to estimate the specific heat capacity of graphene nanofluids, we propose that the density of graphene is also close to the density of graphite, which is 2200 kg/m^3. To calculate the thermal efficiency η_t in the case of side irradiation, we assume *F* to be equal to half of the lateral surface area of the cylindrical vessel, which is 0.007 m^2.

$$C_{nf} = \frac{\phi\rho_{np}C_{np}+(1-\phi)\rho_{bf}C_{bf}}{\phi\cdot\rho_{np}+(1-\phi)\cdot\rho_{bf}} \quad (6)$$

where the indices *nf* refer to the nanofluid, *np* to the nanoparticles, and *bf* to the base fluid; ρ represents density, and ϕ denotes the volume concentration of nanoparticles.

The calculated results of the thermal efficiency η_t for graphene nanofluids with different volume concentrations are presented in Table 1. Through careful analysis, it has been ascertained that the application of rotating graphene nanofluid in this investigation yielded a 2.5% enhancement in thermal efficiency compared to distilled water, with a graphene volume concentration of 0.1%. Nevertheless, as the graphene concentration was further increased, a reduction in efficiency was observed. This decline can be attributed to the degradation of the optical and thermophysical characteristics of the graphene nanofluid due to coagulation and sedimentation phenomena. Additionally, it is important to highlight that the overall efficiency of converting solar radiation into thermal energy can be improved through proper system design and the implementation of measures aimed at minimizing heat losses to the surrounding environment.

Table 1. The results of calculating the thermal efficiency for the case of side irradiation

Parameters	Volume concentration of graphene ϕ, %			
	0	0.1	0.5	1.0
$T_m^b - T_0$, °C	51.4	52.1	53.2	53.1
Δm, g	5.01	5.46	5.20	4.17
c_p, J/(kg·K)	4182.5	4175.0	4145.3	4108.5
η_t	0.669	0.694	0.688	0.639

4 Conclusion

In this study, for the first time, an investigation was conducted to study the heating and evaporation process of graphene nanofluid in a rotating volume heated by a solar radiation simulator. Based on the obtained results, the following conclusions can be drawn:

- The evaporation rate of water strongly depends on the irradiation direction, concentration of graphene nanoflakes, and rotation speed. When irradiated laterally, the evaporation rate increases with increasing concentration of graphene nanoflakes, reaching its maximum value at a volume concentration of 0.1%, and then decreases as the concentration increases to 0.5% and 1.0%. In the case of top irradiation, the graphene nanofluid evaporates slower than distilled water at all concentrations. It has been shown that the evaporation rate monotonically increases with increasing flow rotation speed.
- The results of measuring the temperature field of the interfacial surface between fluid and air showed significant differences between side and top irradiation. It has been demonstrated that the graphene film on the surface and the coagulated graphene are heated less compared to the base fluid in the case of side irradiation. However, in the case of top irradiation, the graphene film is heated to a high temperature of approximately 140°C.
- It has been shown that the forced rotation of the graphene nanofluid enhances the process of graphene nanoflake coagulation. Furthermore, the formation of a graphene film on the interfacial surface also reduces the overall evaporation surface area. These factors significantly affect the efficiency of the heating and evaporation process. Addressing these issues can lead to more noticeable improvements in heating and evaporation of graphene nanofluid.
- Equations have been derived for calculating the profile and surface area of the interfacial surface of a rotating liquid. The results of evaluating the efficiency of heating and evaporation in rotating graphene nanofluid have demonstrated that the maximum improvement (2.5%) compared to distilled water was achieved for a sample with a volumetric graphene concentration of 0.1%.

- There are several issues and promising research areas related to the application of graphene in the field of energy that can be highlighted. For instance, investigating methods to enhance the stability of graphene-based nanofluids; studying the interaction mechanism between graphene and solar radiation, considering their orientation and number of layers; exploring the heat transfer mechanism from the graphene surface to the base fluid, as well as the nano- and micro-bubble generation mechanism near the graphene surface during boiling. Additionally, the development of novel graphene-based structures for solar-driven clean water generation.

The authors would like to acknowledge the financial support from the Russian Science Foundation, project 23-19-00840.

References

1. A.S. Dmitriev, A.V. Klimenko, Therm. Eng., **67**, 77–91 (2020)
2. A.S. Dmitriev, A.V. Klimenko, Therm. Eng., **70**, 551–572 (2023)
3. Q.T. Tran, A.S. Dmitriev, P.G. Makarov, I.A. Mikhailova, COLLOID J+, **85**, 837–848 (2023)
4. A. Svobodova-Sedlackova; A. Calderón; C. Barreneche; R. Salgado-Pizarro; P. Gamallo; A.I. Fernández, J. Nanofluids, **12**, 157–172 (2023)
5. E.C. Okonkwo, I. Wole-Osho, I.W. Almanassra, Y.M. Abdullatif, T. Al-Ansari, J. Therm. Anal. Calorim., **145**, 1–56 (2020)
6. O. Mahian, E. Bellos, C.N. Markides, R.A. Taylor et al., Nano Energy, **86**, 106069 (2021)
7. A. Kumar, A.K. Tiwari, Z. Said, Sustain. Energy Technol. Assess., **47**, 101417 (2021)
8. M. Awais, A.A. Bhuiyan, S. Salehin, M.M. Ehsan et al., Int. J. Thermophys., **10**, 100086 (2021)
9. J. Traciak, J. Sobczak, R. Kuzioła et al., J. Therm. Anal. Calorim., **147**, 7665–7673 (2022)
10. M.U. Sajid, Y. Bicer, Sol. Energy, **207**, 974–1001 (2020)
11. H. Liu, Z. Huang, K. Liu, X. Hu, J. Zhou, Adv. Energy Mater., **9**, 1900310 (2019)
12. E.A. Arkhipova et al., J. Therm. Anal. Calorim., **140**, 2641–2648 (2020)

Parametric Analysis of a Latent Heat Storage System for Predicting the Charging and Discharging Time

John Nico Omlang[1*], and *Jaime* Honra[1]

[1]School of Mechanical, Manufacturing, and Energy Engineering, Mapua University, Intramuros, Manila, 1002 Metro Manila, Philippines

Abstract. Latent Heat Storages (LHS) are essential for harnessing renewable energy by storing surplus energy for later use. This research aims to enhance the predictability of charging and discharging times in LHS systems, a crucial step for effective energy management and optimizing renewable energy utilization. A comprehensive parametric analysis was conducted using computational fluid dynamics (CFD) to develop correlation equations for forecasting these times under diverse geometric and operational settings. Parameters considered includes the heat transfer fluid (HTF) inlet temperature, phase change material (PCM) thickness-to-length ratio, Reynold's number, and HTF inlet location. The enthalpy-porosity method and Boussinesq model were employed in numerical simulations to account for natural convection during charging. The influence of the mushy-zone constant on the PCM's temperature profile was investigated and found to be significant. The bottom HTF injection was found to reduce the charging times, while extending discharging durations. The thickness-to-length ratio of the PCM emerged as the most influential factor, with Reynold's number exerting the least influence in the cycle. Analysis showed that increasing PCM thickness-to-length ratio consistently led to longer charging time. Four correlations were developed with an impressive average multiple R of 0.999463, signifying high predictability for charging and discharging times.

1 Introduction

In recent years, phase change materials (PCMs) have gained industry interest due to their high thermal energy storage capacity, offering a solution to address energy supply and demand mismatches. PCMs store energy when available and release it when needed. Numerical simulations, classified as pure conduction (PC) or combined conduction and natural convection (CCNC), are used to analyze PCM charging in latent heat storage (LHS) systems, reducing experimental costs. In 2015, Seddegh et al. examined a vertical shell-and-tube LHS, employing PC and CCNC models for heat transfer analysis. Their research validated the numerical simulations with experimental data, revealing that the CCNC model

* Corresponding author: jnomlang@gmail.com

better captures energy transfer during charging, while the PC model is more suitable for discharging. Interestingly, no significant difference in discharging time was observed between the two models [1]. In 2017, a triplex tube heat exchanger with PCM was studied, investigating simultaneous charging and discharging. They used Paraffin RT82 for charging and RT60 for discharging, with two heating modes. The study employed PC and CNCC models, finding that upward liquid PCM motion influenced charging and recommending the PC model for discharging due to simulation errors with CNCC [2]. In 2020, a study explored simultaneous PCM charging and discharging using microencapsulated PCM mixed with carbon fibers to prevent leakage. They analyzed temperature, energy storage, and different flow rates. Interestingly, initially melted PCM yielded higher stable temperatures [3]. Fornarelli et al. (2017) used 2D simulations to study a shell-and-tube LHS with molten salt PCM and diathermic oil. They observed a curved phase change interface during charging and noted the absence of convective motion in discharging [4]. In 2020, an investigation of the influence of side walls in thermal energy storage was performed and found that straight walls were best for granular PCM in LHS [5]. Dixit (2020) predicted LHS performance through a correlation analysis, varying inlet conditions, and storage volume, concluding that 2D simulations suffice for modelling PCM charging in vertical annular cylinders [6].

This study aims to enhance the predictability of charging and discharging times in latent heat storage systems, vital for efficient renewable energy utilization. These systems store excess energy for later use but are influenced by various parameters. The main objective of this study is to conduct a parametric analysis on latent heat storage and predict the performance under different operating conditions and variation in geometrical parameters through numerical modelling. While previous studies focused on limited parameters, this research conducts a comprehensive parametric analysis using numerical simulations. It assesses the impact of factors like heat transfer fluid (HTF) inlet temperature, PCM thickness, LHS device length, Reynolds number, and HTF inlet location. The study also develops dimensionless correlations to predict LHS cycle performance accurately, facilitating efficient integration with diverse energy sources.

This study holds significance by introducing additional parameters that provide valuable insights into the charging and discharging cycles of LHS. The correlations developed here can serve as a reference for predicting PCM sizing and performance in heat exchangers, considering geometrical parameters and initial conditions of HTF and PCM. This research offers potential benefits for energy companies investing in renewables, enhancing efficiency and storage capacity, thus reducing costs, and increasing the viability of solar and wind power. Additionally, it applies to building design with latent heat storage, leading to reduced energy consumption, lower operational costs, and sustainable construction. Policymakers can also utilize these findings to promote efficient, sustainable energy storage, contributing to a low-carbon economy and reduced environmental impact. Furthermore, the study's modelling and simulation techniques have broader applications in energy storage and sustainable systems, enhancing accuracy and reliability across various fields.

2 Numerical modelling

The simulations conducted in this study adopted a 2D axisymmetric approach to minimize computational cost. The chosen viscous model for these simulations is laminar, aligning with prior validation studies that exclusively examined the laminar regime for HTF flow. In a parallel manner, this study was designed considering the laminar flow of the HTF and the laminar regime associated in the melting process of the PCM.

While natural convection must be taken into account when analyzing the solidification and melting of materials inside a container, the enthalpy-porosity approach is frequently used. Due to its quick convergence and great accuracy advantage over other approaches, this is commonly employed. In the study conducted in 2009 [7], a successful validation of the enthalpy-porosity method was done against experimental results [8].

The solidification and melting process is modeled in ANSYS Fluent using the enthalpy-porosity method. In this method, the melt contact is not specifically recorded. Instead, a melt percentage is connected to each cell in the domain. The melt fraction is computed using an enthalpy balance at each iteration. An area between 0 and 1 in terms of melt fraction is referred to as the "mushy zone." A pseudoporous medium with a porosity of one that drops to zero as the substance solidifies is used to represent the mushy zone. When a substance has fully solidified inside a cell, its porosity is zero, and hence, its velocities are also zero [9]. Equation 1 shows the enthalpy of a material which is the sum of sensible, h, and latent heat, ΔH.

$$H = h + \Delta H \tag{1}$$

where:

$$h = h_{ref} + \int_{T_{ref}}^{T} C_p dT \tag{2}$$

h_{ref} = reference enthalpy
T_{ref} = reference temperature
C_p = specific heat at constant pressure

The Equation 3 illustrates how the latent heat content can be expressed in terms of the liquid percentage, β, and the material's latent heat, L. For a solid or liquid, the latent heat constant ranges from zero to L, respectively [9].

$$\Delta H = \beta L \tag{3}$$

Where β is defined as:

$$\beta = \begin{cases} 0 & if \;\; T < T_{solidus} \\ 1 & if \;\; T > T_{solidus} \\ \frac{T - T_{solidus}}{T_{liquidus} - T_{solidus}} & if \;\; T_{solidus} < T < T_{liquidus} \end{cases} \tag{4}$$

The energy equation involved in melting or solidification problems is defined as:

$$\frac{\partial}{\partial t}(\rho H) + \nabla \cdot (\rho \vec{v} H) = \nabla \cdot (k \nabla T) + S \tag{5}$$

where:
H = enthalpy
ρ = density
$\vec{v}$ = velocity of fluid
S = source term

The temperature problem is just an iterative process between Equations 4 and 5. The energy equation typically does not converge well when directly updating the liquid fraction using Equation 4 [9].

The mushy zone is regarded as a porous material by the enthalpy-porosity method. Each cell's porosity is set to be equal to its liquid fraction. The porosity is zero in fully solidified regions which results in zero velocity in those areas. The momentum sink caused by the mushy zone's reduced porosity is given by Equation 6 [9].

$$S = \frac{(1-\beta)^2}{(\beta^3+\varepsilon)} A_{mush}(\vec{v} - \vec{v}_p) \quad (6)$$

where:
β = liquid fraction
ε = a small value (0.001), preventing division by zero
A_{mush} = mushy zone constant
$\vec{v}_p$ = pull velocity

The simulation of fluid flow through a porous media in the mushy zone is done by adding a damping factor to the momentum equation through the mushy zone constant. The steeper the shift of the material's velocity to zero when it solidifies, the higher this number. Extremely high values could make the solution oscillate [9-10].

The PCM structure may be impacted by phase-to-phase density changes, although not so much the heat transmission. When there are fluctuations in density inside a liquid or between phases, different strategies are used. The Boussinesq approximation, which is described by Equation 7, is one method. Recent research has extensively employed this approximation method to examine how natural convection affects the thermal performance of PCMs. Only the buoyancy term of the momentum is taken into account in this method when calculating the PCM's change in density [8].

$$\rho = \rho_0(1 - \beta \Delta T) \quad (7)$$

where:
ΔT = cell temperature and reference temperature (T_0) difference
ρ_0 = reference density at temperature T_0

In modeling of phase change problem using enthalpy-porosity method, the following are the usual assumptions [8]:

- Fluid flow is transient.
- Gravitational force is considered.
- The flow of liquid PCM is laminar due to low Reynold's number.
- The liquid PCM is incompressible.
- The solid and liquid phase of PCM is homogeneous and isotropic.
- The thermophysical properties of PCM are temperature dependent.
- The density of PCM only changes during phase transition.
- Radiation heat transfer is ignored.
- Thermal resistances between the solid and liquid phase of the PCM are negligible.
- Surface tension effects are neglected.

When considering the robustness and computational efficiency of the model, it is imperative to meticulously select appropriate under-relaxation factor (URF) values. These are numerical parameters used to control the convergence of iterative solvers. The significance of this parameter becomes particularly pronounced when employing the enthalpy-porosity Volume of Fluid (VOF) approach in numerical simulations. To facilitate this selection process, Table 1 provides a comprehensive list of suggested URFs tailored for various numerical cases [8].

Moreover, achieving a reliable numerical solution for phase change problems requires the cautious application of both time and spatial discretization schemes. These schemes play a pivotal role in ensuring the accuracy and stability of the simulation. Delving into this aspect, Table 2 presents a compilation of recommended schemes that can be effectively utilized in tandem with the enthalpy-porosity VOF approach. This dual consideration of URF values and discretization schemes serves as a foundational step toward enhancing the overall precision and computational efficiency of the model, thereby contributing to the reliability of the results obtained [8].

Table 1. URF for PCM Modeling Using Enthalpy-Porosity Method

URF	Value/s
Pressure	0.3
Momentum	0.1 – 0.3
Density	0.7
Volume fraction	0.5
Liquid fraction	0.95
Energy	0.95

Table 2. Discretization Schemes for PCM Modeling Using Enthalpy-Porosity Method

Transient formulation	Bounded second order implicit
Pressure-velocity coupling	PISO/Coupled
Spatial discretization	
Gradient	Least squares cell based
Pressure	Presto!
Density	Quick
Momentum	Quick
Volume fraction	Compressive
Energy	Quick

3 Numerical model validation

To ensure the credibility of the numerical model employed in both the charging and discharging processes of the LHS device, a rigorous validation process was conducted. This validation involved a comparison of the model's outcomes with both experimental data from a study by Longeon et al. (2013) and the numerical model developed in a previous investigation by Dixit (2020). For the comparative analysis, a specific point, denoted as point X, was selected. This point is positioned at 6 mm from the outer surface of the pipe and 217 mm from the bottom of the system. A temperature profile was generated at this point to serve as the basis for comprehensive validation through comparison with the experimental results and numerical model from previous study. The geometries were modeled directly using the built-in design modeler of ANSYS Fluent 2019. These geometries represented the computational domain for the CFD analysis. Figure 1 shows the geometry from the study of Longeon et al. (2013) that was used for the numerical model validation [11]. In the simulation of the charging process, the average velocity of the HTF inlet was set to 0.01 m/s at 52°C while the outlet was set to pressure outlet at 0 kPa gage. The PCM's initial temperature was set to 22°C with the outer surface being adiabatic. The same boundary conditions were applied to simulation of the discharging process except that the HTF inlet temperature was set to 17°C and the initial temperature of the PCM was at 46°C. Both simulations used top injection of the HTF.

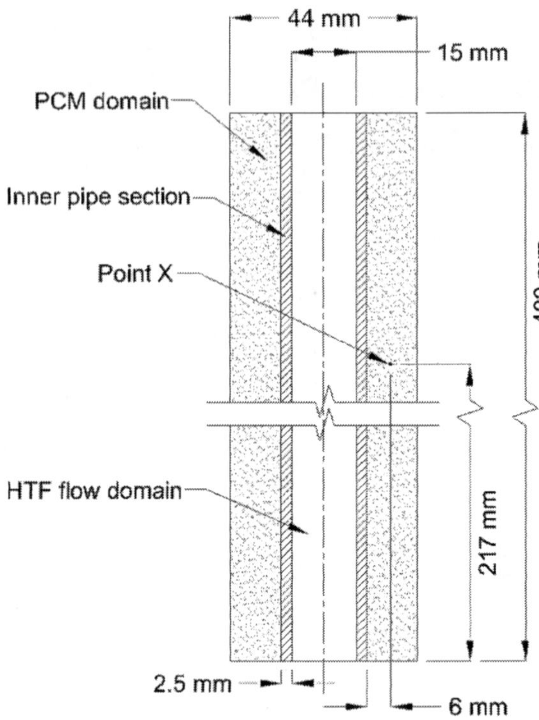

Fig. 1. Geometry specification for numerical model validation.

The validation process includes assessment of the numerical model's response to variations in the mushy-zone constant. Additionally, an in-depth evaluation was conducted to gauge the sensitivity of the model to changes in both time step size and grid size. These thorough analyses ensure the reliability and accuracy of the numerical model, providing valuable insights into its performance across various conditions and parameters.

3.1 Influence of mushy-zone constant

The influence of mushy-zone constant on the charging and discharging of LHS was investigated in this study. The model was simulated under mushy-zone constants of 1×10^5, 1×10^6, and 1×10^7. The result shows that the mushy-zone constant does not have significant effect on the liquid fraction profile of the PCM during the charging and discharging process as shown in Figure 2 and Figure 3, respectively, although there are slight differences on the total charging time. On the contrary, a substantial effect was observed in the temperature profile of the selected point X during the charging process of the LHS as illustrated in Figure 4. The effect in the temperature profile of point X during discharging, however, remains insignificant which is clearly presented in Figure 5.

 https://doi.org/10.1051/e3sconf/202451403006

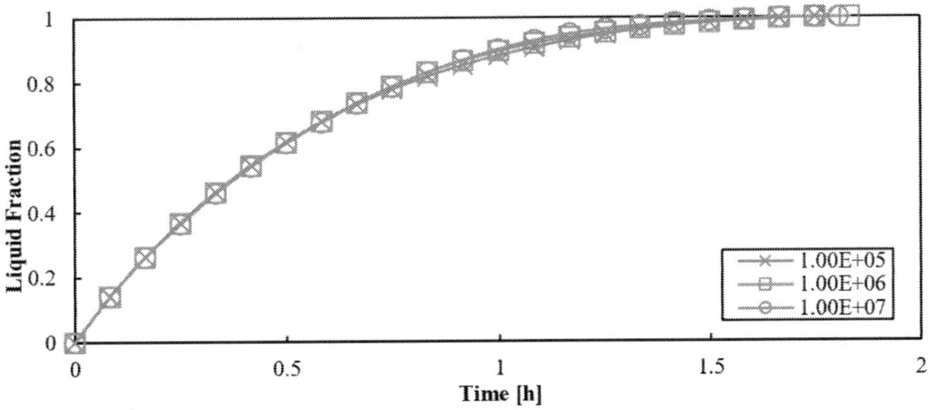

Fig. 2. Liquid fraction profile of PCM domain under varying mushy-zone constant during charging process.

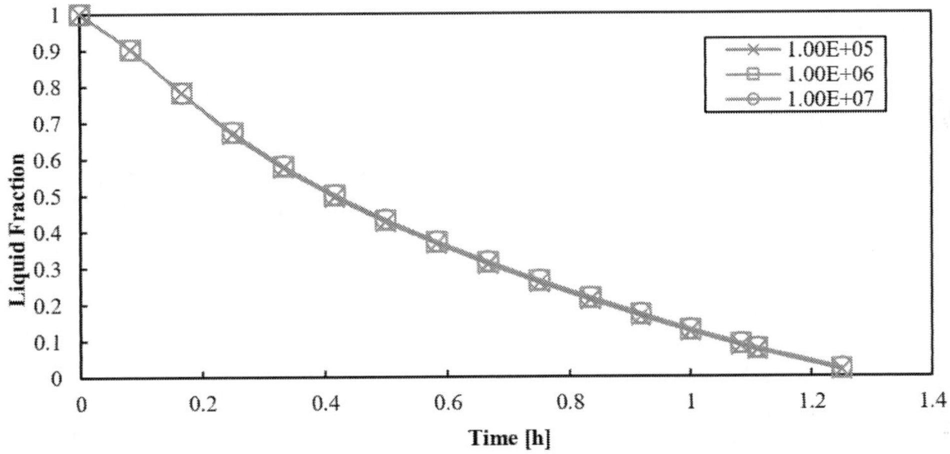

Fig. 3. Liquid fraction profile of PCM domain under varying mushy-zone constant during discharging process.

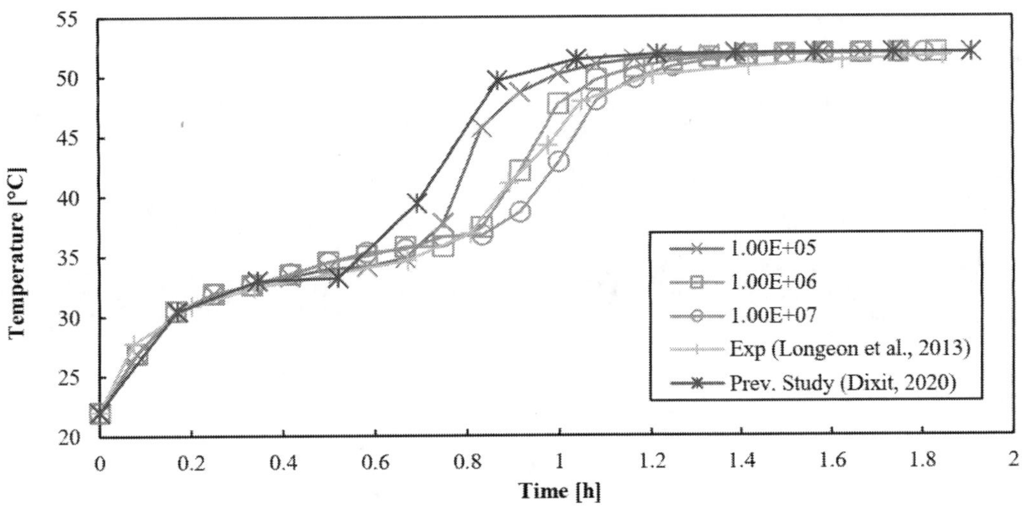

Fig. 4. Temperature profile at point X under varying mushy-zone constant during charging process.

https://doi.org/10.1051/e3sconf/202451403006

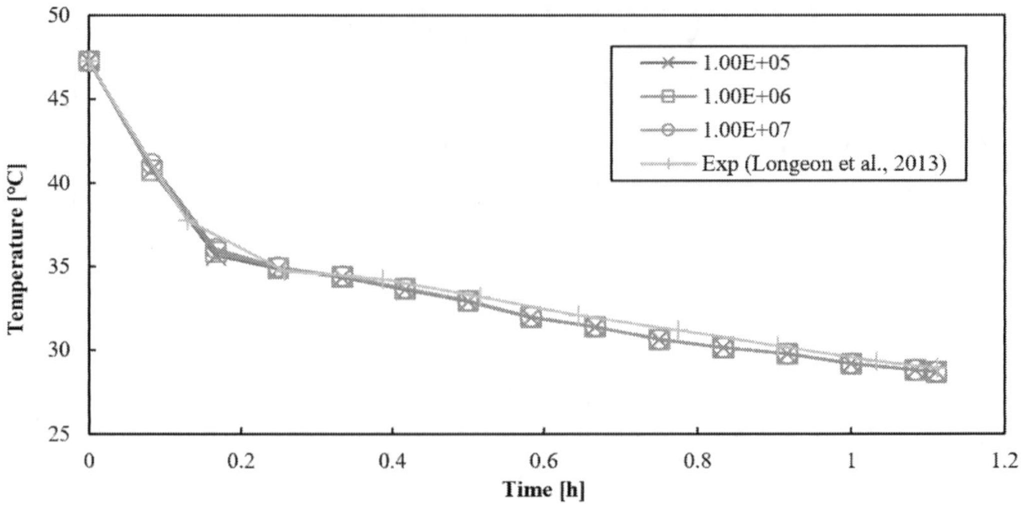

Fig. 5. Temperature profile at point X under varying mushy-zone constant during discharging process.

3.2 Time and grid independence test

In this study, a comprehensive analysis was conducted to evaluate the impact of various factors on the charging and discharging processes of an LHS. Considering geometric variations, maintaining stable simulation results across changes in time step and grid size is crucial. To assess the numerical model's sensitivity to time and mesh parameters, the time step size was systematically varied from 0.01 second to 1 second, and the grid size was adjusted from coarse to fine in terms of cell count. Two-dimensional quadrilateral cells are used with average skewness of zero, as shown in Figure 6. Figure 7 illustrates the model's sensitivity between 1,900 cells and 4,875 cells. Notably, increasing cells beyond 4,875 showed no further changes in melting time. Thus, a minimum of 4,875 cells was deemed sufficient for simulations, striking a balance between accuracy and computational efficiency. Furthermore, the model underwent testing for time sensitivity while maintaining a cell count of 4,875. As shown in Figure 8, variations in the time step size did not affect the total melting time. This finding suggests that a 1-second time step size can be employed to expedite computations, though with an increased number of maximum iterations per time step to ensure convergence to the predefined criteria.

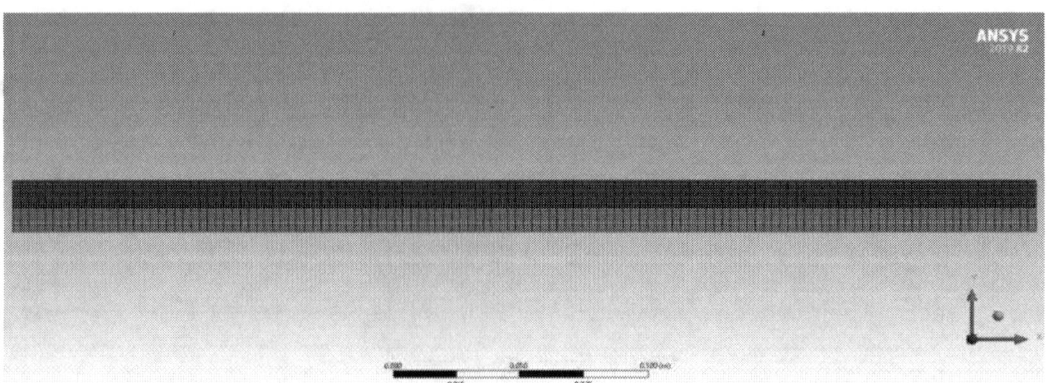

Fig. 6. 2D quadrilateral mesh of the domain.

https://doi.org/10.1051/e3sconf/202451403006

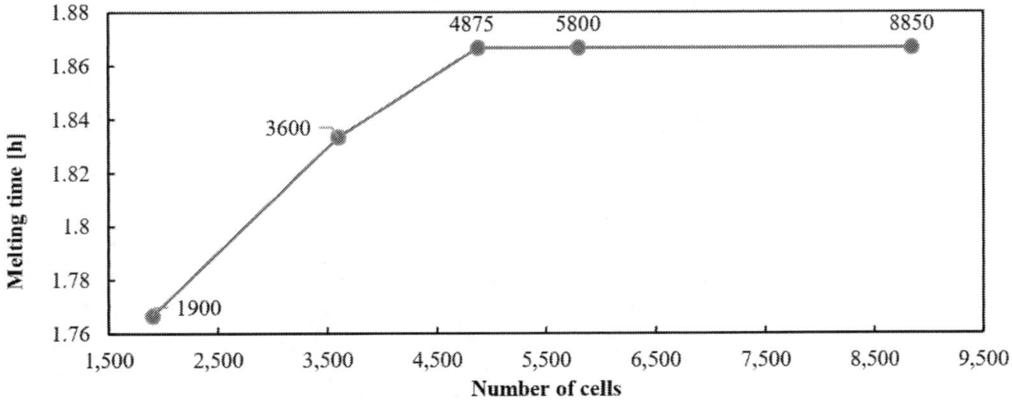

Fig. 7. Grid size effect on PCM melting time.

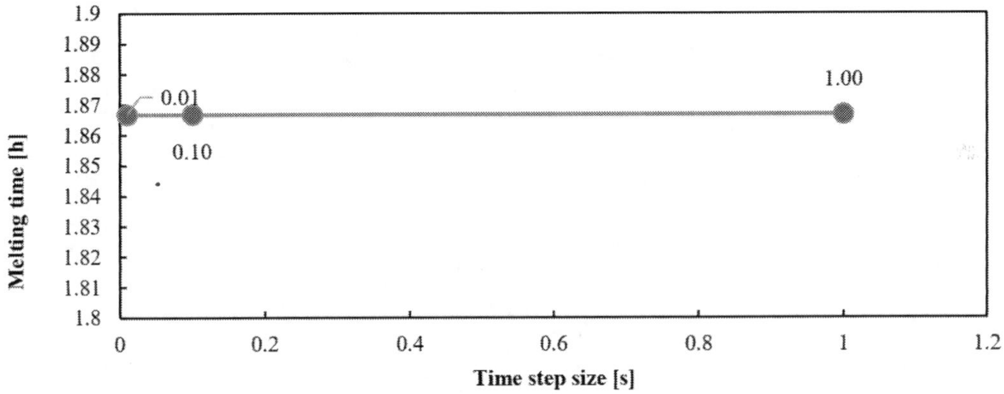

Fig. 8. Time step size effect on PCM melting time.

4 Results and discussion

4.1 Parametric analysis

The parameters included in the study are HTF inlet location, HTF inlet temperature, Reynold's number, and the PCM thickness-to-length ratio. Each parameter was varied, and the total melting time were calculated using the numerical model developed. It is noticeable that during the charging process of the LHS, the injection of HTF at the bottom resulted in shorter charging times compared to that with top injection. The percent reduction in charging time for every case is shown in Table 3. It is also evident that increasing the PCM thickness-to-length ratio leads to a noticeable increase in charging time, regardless of the other factors.

Table 3. Charging time comparison between top and bottom HTF injection.

HTF temperature (°C)	Reynold's number	PCM t/L ratio	Top HTF injection charging time (h)	Bottom HTF injection charging time (h)	Decrease in charging time
45	1000	0.015	0.5333	0.4500	15.6%
		0.02	2.4333	2.3167	4.8%

		0.0225	5.4500	5.2667	3.4%
	1500	0.015	0.5333	0.4500	15.6%
		0.02	2.4000	2.3000	4.2%
		0.0225	5.3667	5.2167	2.8%
	2000	0.015	0.5167	0.4583	11.3%
		0.02	2.3833	2.2833	4.2%
		0.0225	5.3167	5.1667	2.8%
55	1000	0.015	0.3333	0.2833	15.0%
		0.02	1.4167	1.3500	4.7%
		0.0225	3.0333	2.9167	3.8%
	1500	0.015	0.3333	0.2833	15.0%
		0.02	1.4000	1.3333	4.8%
		0.0225	2.9833	2.8833	3.4%
	2000	0.015	0.3167	0.2833	10.5%
		0.02	1.3833	1.3333	3.6%
		0.0225	2.9500	2.8500	3.4%
65	1000	0.015	0.2667	0.2167	18.7%
		0.02	1.0167	0.9667	4.9%
		0.0225	2.1500	2.0500	4.7%
	1500	0.015	0.2500	0.2167	13.3%
		0.02	1.0000	0.9667	3.3%
		0.0225	2.1167	2.0167	4.7%
	2000	0.015	0.2500	0.2167	13.3%
		0.02	1.0000	0.9500	5.0%
		0.0225	2.0833	2.0000	4.0%

On the other hand, the bottom injection of the HTF during the discharging process resulted in longer discharging times compared to that with top HTF injection. However, there are no obvious relationships between the parameters and the amount of increased discharging time. The percent increase in discharging times for each case are presented in Table 4.

Table 4. Discharging time comparison between top and bottom HTF injection.

HTF temperature (°C)	Reynold's number	PCM t/L ratio	Top HTF injection discharging time (h)	Bottom HTF injection discharging time (h)	Increase in discharging time
22	1000	0.015	0.4333	0.4667	7.7%
		0.02	1.8833	2.0667	9.7%
		0.0225	4.7833	5.1167	7.0%
	1500	0.015	0.4167	0.4667	12.0%

https://doi.org/10.1051/e3sconf/202451403006

		0.02	1.8833	2.0500	8.9%
		0.0225	4.7833	5.1000	6.6%
	2000	0.015	0.4167	0.4500	8.0%
		0.02	1.8833	2.0333	8.0%
		0.0225	4.7667	5.0833	6.6%
12	1000	0.015	0.2667	0.2778	4.2%
		0.02	1.1000	1.2000	9.1%
		0.0225	2.7500	2.9500	7.3%
	1500	0.015	0.2500	0.2667	6.7%
		0.02	1.1000	1.2000	9.1%
		0.0225	2.8000	2.9333	4.8%
	2000	0.015	0.2500	0.2667	6.7%
		0.02	1.1000	1.1833	7.6%
		0.0225	2.7500	2.9333	6.7%
2	1000	0.015	0.1833	0.2000	9.1%
		0.02	0.8000	0.8667	8.3%
		0.0225	1.9833	2.1167	6.7%
	1500	0.015	0.1833	0.2000	9.1%
		0.02	0.8000	0.8667	8.3%
		0.0225	1.9833	2.1167	6.7%
	2000	0.015	0.1833	0.2000	9.1%
		0.02	0.8000	0.8667	8.3%
		0.0225	1.9667	2.1000	6.8%

In order to further analyze the impact of each parameter on the charging and discharging processes of the LHS, a comprehensive analysis was conducted utilizing the Pareto principle as a guiding framework. As illustrated in Figure 9 and Figure 10 for charging and discharging, respectively, Pareto charts were constructed to visually represent the significance of the parameters. In adherence to the 80/20 rule, it was strikingly evident that only the thickness-to-length ratio of the PCM played a pivotal role in influencing both the charging and discharging process of the LHS device. This singled out the HTF temperature, HTF inlet location, and the Reynold's number as factors of relatively minor influence in both charging and discharging process. Moreover, the results from the analysis clearly showed that the Reynold's number didn't exert much influence on the charging and discharging times of the LHS. This finding provides an important insight into the hierarchy of factors that affect the LHS device's performance, offering valuable guidance for optimizing its operation.

https://doi.org/10.1051/e3sconf/202451403006

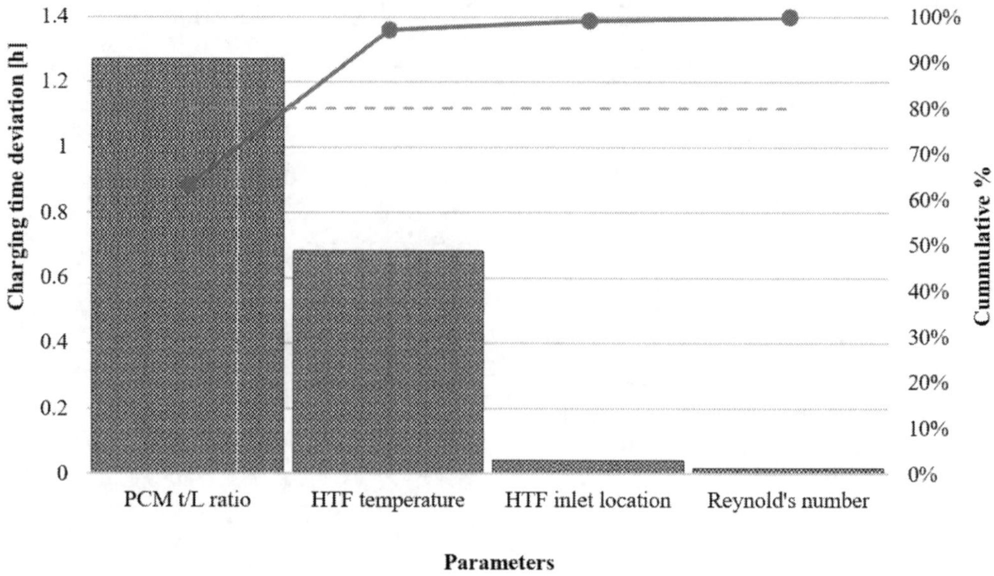

Fig. 9. Pareto chart: Influence of parameters on charging time of LHS.

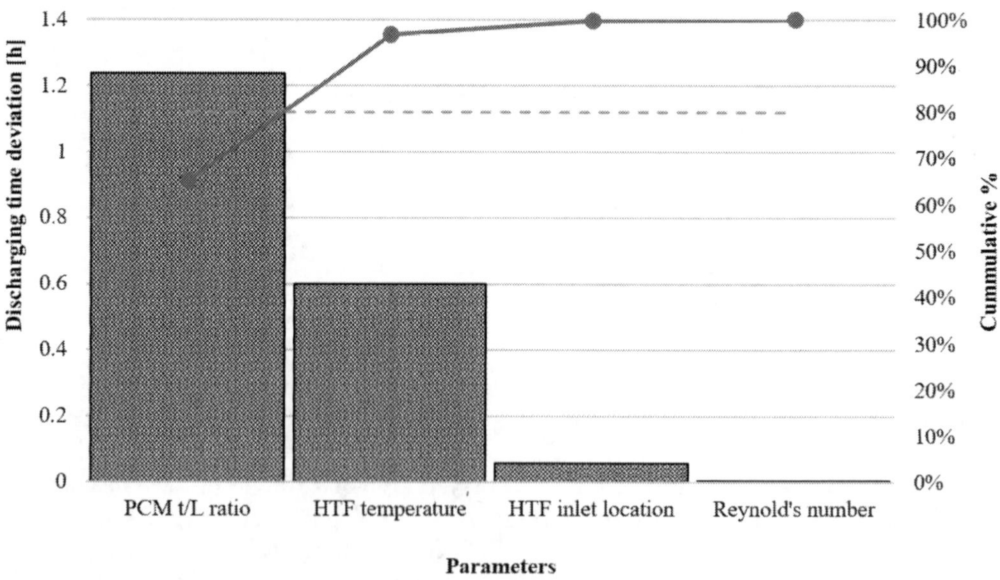

Fig. 10. Pareto chart: Influence of parameters on discharging time of LHS.

4.2 Development of non-dimensional correlation

The correlation was developed using power law equation where dimensionless numbers were used. The Reynold's number was used as one of the non-dimensional parameters of the correlation, which represents the variation of velocity of HTF. The Grashof number, which is given by Equation 8, was considered to represent the variation in the inlet temperature of the HTF and the thermophysical properties of the PCM used. β represents the coefficient of thermal expansion of PCM, t_{PCM} is the thickness of the PCM, T_i is the inlet temperature of HTF, T_{melt} is the melting temperature of PCM, and v is the kinematic viscosity of PCM.

$$Gr = \frac{g\beta(2t_{PCM})^3(T_i - T_{melt})}{v^2} \quad (8)$$

The cross-sectional area ratio, defined as the ratio of the cross-sectional area of PCM and cross-sectional area of HTF, represents the cross-section variation in the LHS device. It is defined using Equation 9, where $(D_o)_{PCM}$ is the outside diameter of the PCM, $(D_i)_{PCM}$ is the inside diameter of PCM, and $(D)_{HTF}$ is the diameter of the HTF (inside diameter of pipe).

$$A_R = \frac{(D_o^2 - D_i^2)_{PCM}}{(D^2)_{HTF}} \quad (9)$$

The variation in thickness and length of the LHS device was represented by the thickness-to-length ratio of the PCM. It is defined using Equation 10, where t_{PCM} is the thickness of the PCM, and L is the length of the device.

$$\left(\frac{t}{L}\right) = \frac{t_{PCM}}{L} \quad (10)$$

The identified dimensionless numbers were used in the development of the correlation equations. The predicted time for charging and discharging is a function of the Reynold's number, Grashof number, and cross-sectional area ratio as shown in Equation 11. The unknown constants a, b, c, d and e was determined by multiple regression analysis.

$$t = a \times (Re)^b \times (Gr)^c \times (A_R)^d \times \left(\frac{t}{L}\right)^e \quad (11)$$

Equation 11 is written into the form as shown in Equation 12, by taking the logarithm of both sides of the equation.

$$log\, t = log\, a + b\, log(Re) + c\, log(Gr) + d\, log(A_R) + e\, log\left(\frac{t}{L}\right) \quad (12)$$

Using the multiple linear regression model [12], Equation 12 can be written in the form shown in Equation 13 where X_1, X_2, X_3, and X_4 are the independent variables, a_0, a_1, a_2, a_3, and a_4 are the numerical coefficients, and Y is the dependent variable.

$$Y = a_0 + a_1X_1 + a_2X_2 + a_3X_3 + a_4X_4 \quad (13)$$

The regression analysis tool in Microsoft Excel was used to solve for the coefficients of the Equation 13 using 95% confidence level.

For the first correlation, the charging for top HTF inlet is considered. The multiple correlation coefficient (multiple R) of the regression is 0.999298, which indicates that the regression model has high predictability of the dependent variable (melting time). The determined coefficients are summarized in Table 5 where the corresponding P-value are shown. It was evident that P-value for the coefficient of X_1 is greater than the conventional 5% ($P<0.05$) which indicates that the variable X_1 has no significant correlation with Y. Since $X_1 = \log(Re)$, it follows that the Reynold's number has no significant correlation with the charging time for top HTF injection.

Table 5. Regression results for top HTF injection charging.

	Coefficients	**P-value**
Intercept	8.918240593	2.4105×10^{-10}
X_1 (log(Re))	-0.046615779	1.0559×10^{-1}
X_2 (log(Gr))	-0.768157053	5.0818×10^{-23}
X_3 (log(A_R))	2.192818698	8.3528×10^{-15}
X_4 (log(t/L))	3.714287099	8.6790×10^{-9}

The regression analysis was repeated excluding the Reynold's number. The multiple R value for the new model is 0.999207, which still indicates strong predictability of the charging time. The determined coefficients are presented in Table 6.

Table 6. Regression results for top HTF injection charging excluding Reynold's number.

	Coefficients	P-value
Intercept	8.770979462	3.7349×10^{-10}
X_2 (log(Gr))	-0.768157053	2.0871×10^{-23}
X_3 (log(A_R))	2.192818698	7.5362×10^{-15}
X_4 (log(t/L))	3.714287099	1.1801×10^{-8}

From the determined coefficients of the correlation presented in Table 6, the equation for the predicted charging time with top HTF injection (t_{c-top}) in hours is given by Equation 14.

$$t_{c-top} = 59 \times 10^7 (Gr)^{-0.7682} (A_R)^{2.1928} \left(\frac{t}{L}\right)^{3.7143} \tag{14}$$

The same method was used to develop the correlation equations for the predicted charging time with bottom HTF injection ($t_{c-bottom}$), discharging time with top HTF injection (t_{d-top}), and discharging time with bottom HTF injection ($t_{d-bottom}$). For all the correlations, the Reynold's number exhibited no significant correlation with both the charging and discharging time. The summary of developed equations for predicting the charging and discharging are shown in Table 7.

Table 7. Correlation equations developed.

Process	**Injection location**	**Correlation equation**	**Multiple R**
Charging	Top	$t_{c-top} = 59 \times 10^7 (Gr)^{-0.7682} (A_R)^{2.1928} \left(\frac{t}{L}\right)^{3.7143}$	0.999207
	Bottom	$t_{c-bottom} = 3.5245 \times 10^{10} (Gr)^{-0.7753} (A_R)^{2.0266} \left(\frac{t}{L}\right)^{4.667}$	0.999372
Discharging	Top	$t_{d-top} = 4.2355 \times 10^{11} (Gr)^{-1.6861} (A_R)^{4.4626} \left(\frac{t}{L}\right)^{3.3465}$	0.999679
	Bottom	$t_{d-bottom} = 2.2281 \times 10^{12} (Gr)^{-1.6918} (A_R)^{4.3726} \left(\frac{t}{L}\right)^{3.6868}$	0.999593

5 Conclusions

The parametric analysis conducted in this study provided valuable understanding of the factors affecting the performance of an LHS. The correlation between these factors and the charging and discharging time of the device was successfully developed. Noteworthy observations and findings from the analysis include injecting the HTF at the bottom results in faster charging times compared to top injection. Bottom HTF injection during discharging extends discharging times compared to top injection. It was observed that increasing the PCM thickness-to-length ratio consistently leads to a reduced percentage reduction in charging time, irrespective of other factors. The Pareto principle highlights the critical role of PCM thickness-to-length ratio in influencing both charging and discharging processes. Following the Pareto principle, Reynolds number has the least influence on LHS charging and discharging durations. Furthermore, the P-value for Reynolds number coefficients in the regression models exceeds 5%, indicating no significant correlation with charging and discharging times under laminar flow conditions. The four correlation equations exhibit an average multiple correlation coefficient (multiple R) of 0.999463, indicating high predictability of charging and discharging times. It is important to note that the correlation between the parameters and the charging and discharging time holds valid only for the laminar flow condition (Reynold's number from 1000 to 2000) within the range of values considered in the study.

References

1. S. Seddegh, X. Wang, A.D. Henderson, Appl. Therm. Eng., **87**, 698-706 (2015)
2. M. Mastani Joybari, F. Haghighat, S. Seddegh, Energy Build., **139**, 426-438 (2017)
3. Y. Fang, Z.G. Qu, J.F. Zhang, H.T. Xu, G.L. Qi, Appl. Energy, **275**, 115353, (2020)
4. F. Fornarelli, V. Ceglie, B. Fortunato, S.M. Camporeale, M. Torresi, P. Oresta, A. Miliozzi, Energy Procedia, **126**, 501-508 (2017)
5. N.A. Lutsenko, S.S. Fetsov, Renew. Energ., **162**, 466-477 (2020)
6. R. Dixit, University of Massachusetts Amherst Masters Theses. **960** (2020)
7. E. Assis, G. Ziskind, R. Letan, J Heat Transfer, **131(2)**, 1-5 (2009)
8. M. Zeneli, A. Nikolopoulos, S. Karellas, N. Nikolopoulos, Ultra-high temperature thermal energy storage, transfer and conversion, 165-199 (2021)
9. Ansys®, Ansys Fluent Theory Guide, Release 15.0, Help System, Solidification and Melting, ANSYS, Inc. (2013)
10. M. Hameter, H. Walter, Comput. Aided Chem. Eng., **38**, 439-444 (2016)
11. M. Longeon, A. Soupart, J.F. Fourmigué, A. Bruch, P. Marty, Appl. Energy, **112**, 175-184 (2013)
12. M.L. Orlov, Chemistry Department, Oregon State University (1996)

Effects of actuator faults on large-scale wind turbine dynamics with an industry-standard controller

Manh-Hung Do[1*], and *Duc-Duong* Nguyen[2]

[1]Hanoi University of Science and Technology, Hanoi, Vietnam
[2]University of Economics – Technology for Industries, Hanoi, Vietnam

Abstract. The pitch actuator is a critical component of large-scale wind turbines with a high failure rate. Actuator faults significantly impact the dynamics of wind turbines, leading to increased loads on the turbine structure and reduced power output, thereby affecting wind turbine performance and reliability. This paper investigates the effects of actuator faults on a 5-MW reference wind turbine using the OpenFAST simulation software developed by NREL. The fault cases include dynamic changes in the actuator, performance reduction, and, finally, actuator seizing.

1 Introduction

Wind turbine represent sophisticated systems comprising numerous interconnected components. The malfunction of any of these constituent elements precipitates adverse effects on the turbine's functionality, thereby resulting in escalated Operation and Maintenance (O&M) expenditures. Most of the large-scale wind turbines are Horizontal Axis Wind Turbines (HAWT) with three blades. The turbines are equipped with pitch actuators to change the blade attack angle of each blade. The change allows for control of the aerodynamic forces affecting the blades thus controlling the rotor rotational speed as well as loads affecting wind turbine's components. The pitch control system adjusts the pitch angle of the blades to keep the rotor rotating at the rated speed to guarantee the system's safety when the wind speed is above the rated value (fig. 1).

The turbine pitch actuator system has one of the highest failure rates and is the largest contributor to the total downtime of a wind turbine [1,2]. The ReliaWind project reports that the pitch actuator is most responsible for a wind turbine's downtime, accounting for up to 20% of the total failure time [1]. Another survey of wind turbine failures found that 32.25% of the total failures occurred due to the blade angle adjustment system, which accounted for 14.74% of total downtime [2]. Therefore, the study on the impact of pitch actuator faults on wind turbine performance and reliability is of increasing interest.

In case of faults, the behavior of pitch actuators changes leading to aerodynamic imbalances which may result in damage of important components and reduced power production. In literature, the effect of actuator faults is well investigated. In [3], the impact

* Corresponding author: hung.domanh@hust.edu.vn

of the actuator seizing faults on a 10-MW OO-Star semisubmersible platform is discussed. The results showed that the blade seizing fault leads to a significant excitation on the platform motion and the turbine structure. The nacelle acceleration can be increased by 27.9% due to the fault. Dynamic analysis of three types of the spar, semi-submersible, and tension leg floating wind turbines are carried out for pitch actuator seize fault, grid loss, and shut down in [4]. Imbalance loads due to the actuator fault can lead to remarkable loads on the blades and tower. In [5], the effects of stuck blade pitch actuator, fixed value fault, and bias fault are investigated using the HAWC2 simulation tool. The results showed that the pitch sensor stuck is the most severe case affecting the tower, the yaw bearing, and the shaft of wind turbines.

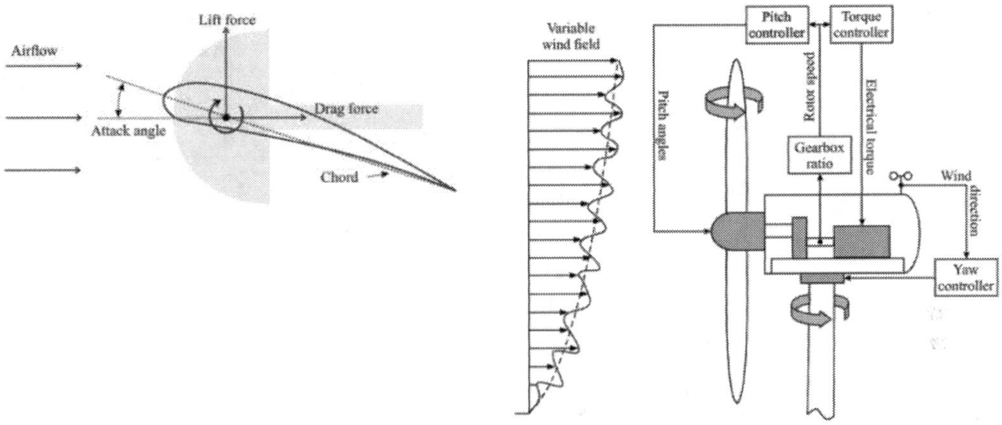

Fig. 1. Wind turbine's pitch angle and control system.

The aforementioned papers mainly study the most severe fault in which the actuator loses function and is stuck at a fixed position. This paper further investigates the most common actuator fault cases including dynamic changes in the actuator, performance reduction, and actuator seizing. Impacts of the faults are analyzed on a 5-MW reference wind turbine [6] using OpenFAST [7] simulation software for both structure loads and power production. To simulate actual turbine operating conditions, TurbSim is used to generate three-dimensional turbulent wind time series in combination with an industry-standard controller.

2 Methodology

2.1 OpenFAST simulation tool

OpenFAST is a physics-based software packet for simulating the coupled nonlinear areo-hydro-servo-elastic response in time domain of wind turbines [7]. OpenFAST includes aerodynamics, hydrodynamics, control and electrical system dynamics, and elastic advanced models. The tool is developed and managed by the National Renewable Energy Lab (NREL). The aerodynamic loads on the blades and tower are calculated using the Blade Element/Momentum (BEM) theory. The blade is divided into many segments. The aerodynamic forces on each segment are then calculated using two-dimensional airfoil data. The two-dimensional flow is approximated from the three-dimensional flow around a body at cross sections. The forces on all of the elements are then summed to find the total load on the blade or tower.

The OpenFAST software is capable of simulating wind turbines with different operating conditions and control algorithms thereby covering the actual operating conditions of a wind turbine. Wind turbine specifications can also be configured in detail, from which users can simulate failure cases on each turbine component. However, OpenFAST does not include the actuator dynamics. Thus, in this paper, OpenFAST is used in conjunction with MATLAB/Simulink to investigate the effects of dynamic changes in the actuator. OpenFAST can be implemented as a dynamic library called from a S-Function in Simulink (fig. 2). The OpenFAST dynamic library responsibles for the nonlinear wind turbine model, actuators and controllers are designed on Simulink. With this configuration, actuator dynamics can be added and modified to simulate fault cases.

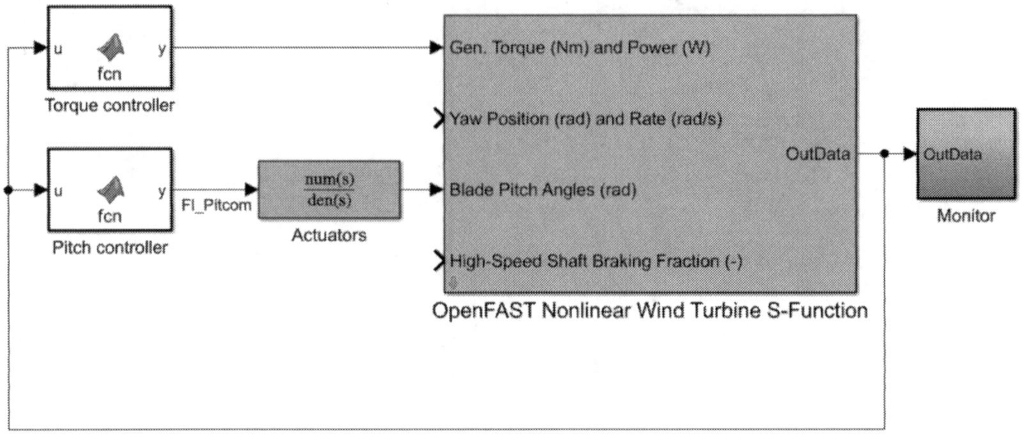

Fig. 2. Simulink and OpenFAST simulation with actuator dynamics.

2.2 Nonlinear reference wind turbine model

An NREL 5-MW baseline wind turbine model [6] is used in this paper. The model is widely used as a reference for investigating and assessing utility-scale multimegawatt wind turbines. The turbine has a three-bladed upwind variable-speed variable-pitch configuration. It was created based on both industrial turbines such as the REpower 5M and some conceptual models from the WindPACT, RECOFF, and DOWEC projects. The detailed properties of the turbine are given in Table 1.

Table 1. NREL 5-MW baseline wind turbine model' properties.

Rating	5 MW
Configuration	Upwind, 3 Blades
Control	Variable Speed, Collective Pitch
Rotor Diameter	126 m
Hub Height	90 m
Cut-In, Rated, Cut-Out Wind Speed	3 m/s, 11.4 m/s, 25 m/s
Cut-In, Rated Rotor Speed	6.9 rpm, 12.1 rpm
Max Power Coefficient	0.483

2.3 Three-dimensional stochastic wind field

To simulate realistic working conditions of wind turbines, three-dimensional stochastic wind fields with different mean wind speed and turbulence intensity (TI) levels are used. The wind fields are generated by TurbSim [8]. The generated wind fields are based on the Kaimal wind turbulence model and follow the IEC 61400-1 (Wind turbines – Part 1: Design requirements) standard. The Kaimal spectrum and exponential coherence model is formulated as:

$$\frac{fS_k(f)}{\sigma_k^2} = \frac{4fL_k/V_h}{(1+\frac{6fL_k}{V_h})^{\frac{5}{3}}}, \quad (1)$$

where f denotes the frequency, k is the velocity component direction, S is single-sided velocity component spectrum, σ is velocity component standard deviation, L is the velocity component integral scale parameter, and V_h is the hub height wind speed.
Examples of wind time series that have a mean speed of 12 m/s and IEC normal turbulence level B with different random seeds are shown in Fig. 3.

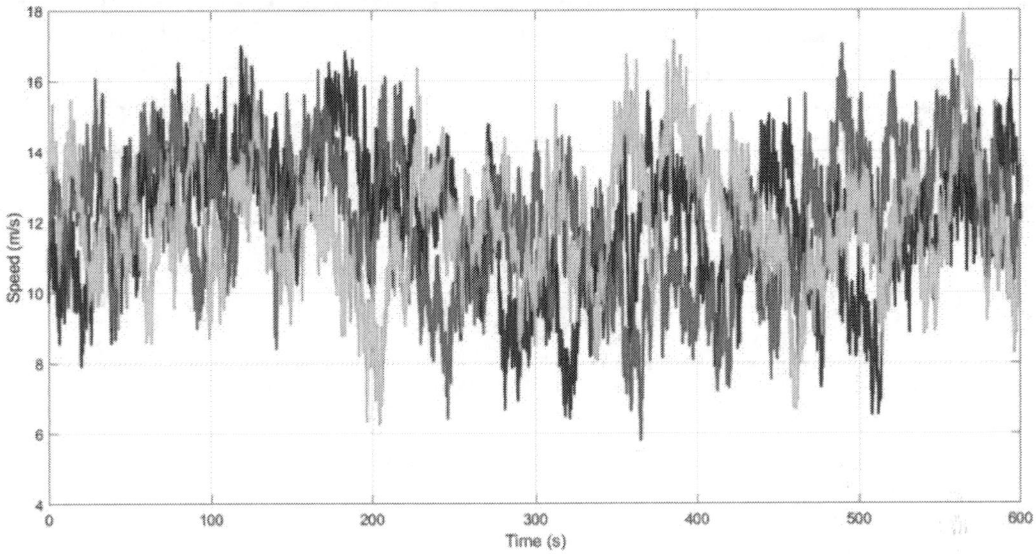

Fig. 3. Hub height stochastic wind speed.

2.4 Control system

The Simulink implementation of NREL's Reference Open Source Controller [9] (ROSCO) is used for both torque and pitch controller in this paper. The controller provides industry-standard functions and performance for the provided wind turbines and thus can be used as a reference to research the impact of actuator faults.

The normal operation of a wind turbine can be divided into four main regions. Region 1 and Region 4 are when the wind speed is outside of the turbine's working region (i.e. below the cut-in and above the cut-out wind speed).

Region 2 is between cut-in and rated wind speed. In this region, the torque controller is employed to maximize the power production, the blade angle is fixed at the optimal value. The power production of a wind turbine strongly depends on the power coefficient C_p which is the ratio of the power produced by the turbine to the theoretical power of the wind at a given wind speed. The power coefficient of a wind turbine is a function of tip-speed ratio (TSR) which is the ratio of rotor and wind speed, and blade pitch angle. If the blade angle is kept at the optimal value, the maximal power production is achieved by tracking the optimal

TSR corresponding to the max C_p. To track the optimal TSR, the generator speed is adjusted following the wind speed as:

$$\omega_{ref} = N_g \frac{\lambda_{opt} v}{R}, \qquad (2)$$

where ω_{ref} is the desired generator speed, N_g is the gearbox ratio of generator and rotor speed, λ_{opt} is the optimal TSR, R is the rotor radius, and v is the wind speed which can be measured or estimated.

The generator speed is controlled following the desired speed by adjusting the generator torque τ_g using a Proportional-Integral (PI) controller:

$$\Delta\tau_g = K_p \Delta\omega_g + K_i \int_0^T \Delta\omega_g \, dt, \qquad (3)$$

where K_p, K_i are the controller gains, $\Delta\omega_g$ is the error between the desired generator speed and the measured generator speed $\Delta\omega_g = \omega_{ref} - \omega_g$.

Region 3 is between rated and cut-out wind speed. Here the generator torque is adjusted to maintain the rated power output, and the pitch controller regulates the rotor at the rated speed. The PI controller is also employed in this region, however, the desired generator speed is constant at the rated value $\omega_{ref} = \omega_{rated}$ and the control input is the blade pitch angle β:

$$\Delta\beta = K_p \Delta\omega_g + K_i \int_0^T \Delta\omega_g \, dt. \qquad (4)$$

Typically, the gains of both torque and pitch PI controller vary based on the wind speed or blade pitch angle using a gain schedule algorithm [10].

2.5 Actuator dynamic

Due to the faster dynamic of the pitch actuator relative to that of the wind turbine, the actuator dynamic is typically considered as a second-order system as:

$$\frac{\beta(s)}{\beta_{com}(s)} = \frac{\omega_n^2}{s^2 + 2\xi\omega_n s + \omega_n^2}, \qquad (5)$$

where $\beta(s)$ is the Laplace transform of the blade pitch angle $\beta(t)$, $\beta_{com}(s)$ is the Laplace transform of the blade pitch angle command $\beta_{com}(t)$, ω_n is the natural frequency and ξ is the damping ratio of the actuator.

When the actuators have initial faults, the dynamic of the actuator changes. The change can be represented by the different natural frequency and damping ratio of the faulty actuator. Typically, the dynamic of a faulty actuator will be slower than that of the fault-free actuator. If the fault continues to develop, the actuator will not be able to pitch the blade to the desired angle. In this case, the fault can be represented by a reduction in the actuator performance as:

$$\frac{\beta(s)}{\beta_{com}(s)} = \eta_a \frac{\omega_n^2}{s^2 + 2\xi\omega_n s + \omega_n^2}, \qquad (6)$$

where η_a denotes the effectiveness of the actuator.

Table 2. Actuator fault cases.

Fault case	Legend	Parameters
Fault free	F0	$\eta_a = 1;\ \omega_n = 6.28\frac{rad}{s};\ \xi = 0.9$
Slow dynamic	F1	$\eta_a = 1;\ \omega_n = 3.14\frac{rad}{s};\ \xi = 0.9$
Performace reduction	F2	$\eta_a = 0.5;\ \omega_n = 3.14\frac{rad}{s};\ \xi = 0.9$
Seized actuator	F3	$\beta(t) = 0.17\ rad$

In the most severe case, the actuator fails to operate so the blade gets stuck at a fixed position $\beta(t)$=const. Table 2 summarizes the actuator fault cases investigated, the step responses of the actuators are given in Fig. 4:

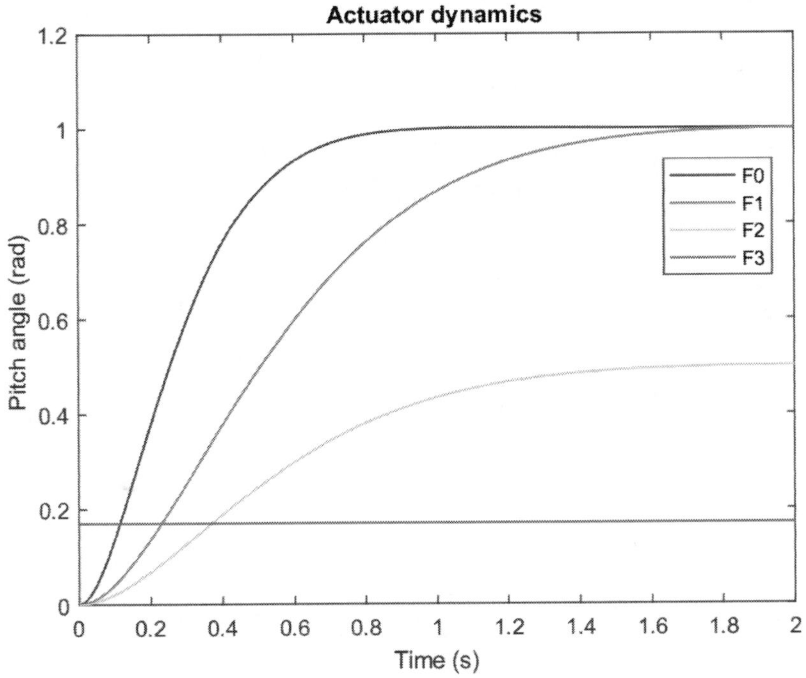

Fig. 4. Step response of actuators (F0: fault-free (blue); F1: Slow dynamic fault (magenta); F2: Reduce performance fault (green), F3: Seized actuator fault (red)).

3 Results and discussions

The simulation results of the NREL 5-MW baseline wind turbine model using OpenFAST software and ROSCO controller for different case studies defined in Table 2 (F0: fault-free (blue); F1: Slow dynamic fault (magenta); F2: Reduce performance fault (green), F3: Seized actuator fault (red)) are shown in this section. It is assumed that there is a fault in the actuator that corresponds to the blade number one of the turbine.

The stochastic, full-field, turbulence wind profile with a mean speed of 12 m/s, and the turbulence level of B according to the IEC standard is used as the input for the simulation (Fig. 5). It can be observed that the dynamics change of the faulty actuator significantly affects the blade pitch angle response. The difference between the pitch angle of the case studies led to the change in power production and structural loads. For the slow actuator dynamics (F1) and reduced performance (F2) fault cases, there is no obvious generator power reduction in comparison with the healthy actuator case. This is because the pitch controller mainly operates in region 3 which is above the rated wind speed. In this region, the goal is not to increase the power production but to regulate the rotor speed a rated value to guarantee the system safety. However, in the seized actuator fault case (F3), the actuator cannot pitch the blade to the optimal angle, the power production has been reduced remarkably.

Fig. 6 illustrates the impact of actuator faults on the low-speed shaft bending moment and tower-top fore-aft deflection of the turbine. To analyze the results, Fig. 7 provides the power spectrums of the structural loads. Due to the aerodynamic imbalances on the blades, the structural loads on the shaft and tower experience an increase. The shaft bending moment experiences an increase mainly in the low-frequency region, while the difference for the tower-top fore-aft deflection is clearly determined at the rated wind speed's 1P frequency (0.2 Hz). The effects of slow actuator fault (F1) are moderately weak since the actuator dynamics are relatively fast compared to the turbine dynamics. However, when the actuator performance reduces (F2) or the actuator seizes (F3), the structural loads increase significantly.

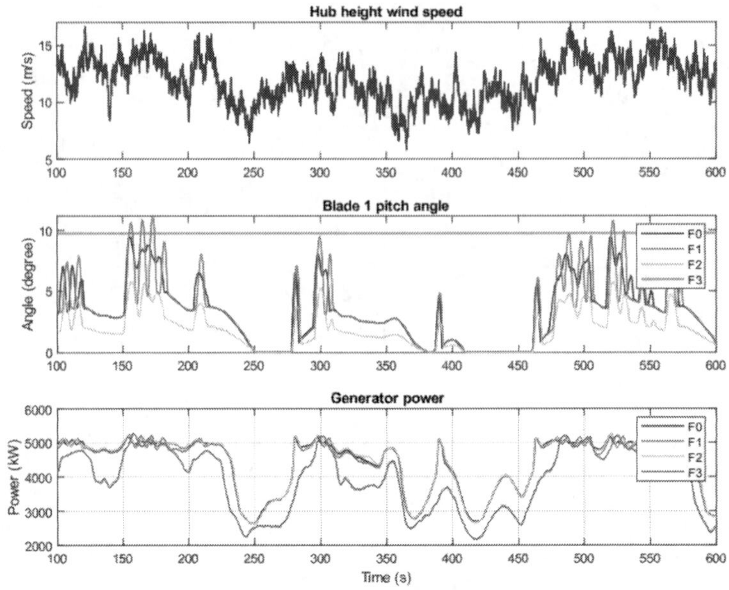

Fig. 5. Wind speed, Blade 1 pitch angle, and Generator power (F0: fault-free (blue); F1: Slow dynamic fault (magenta); F2: Reduce performance fault (green), F3: Seized actuator fault (red)).

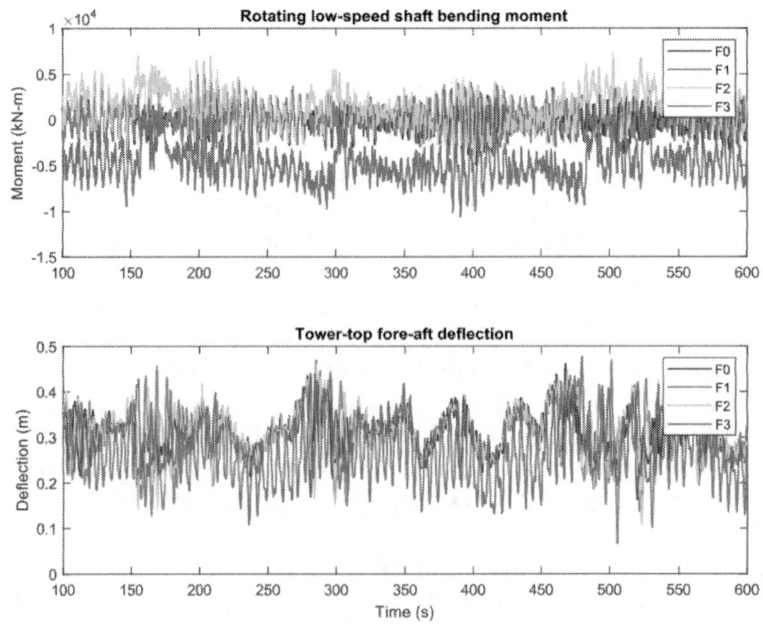

Fig. 6. Rotating rotor shaft bending moment and Tower-top fore-aft deflection (F0: fault-free (blue); F1: Slow dynamic fault (magenta); F2: Reduce performance fault (green), F3: Seized actuator fault (red)).

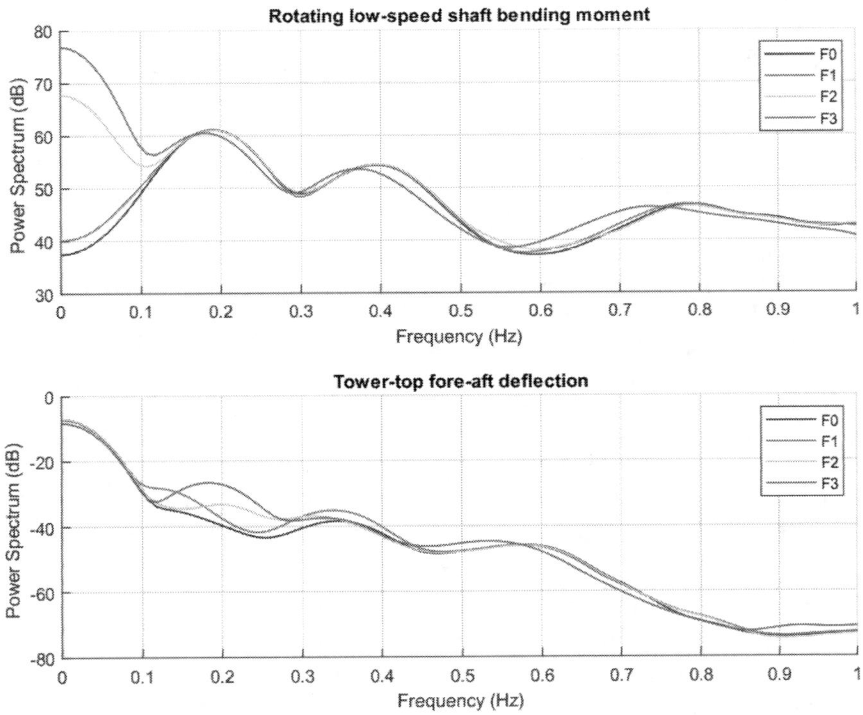

Fig. 7. Power spectrum analysis of Rotating rotor shaft bending moment and Tower-top fore-aft deflection signals (F0: fault-free (blue); F1: Slow dynamic fault (magenta); F2: Reduce performance fault (green), F3: Seized actuator fault (red)).

4 Conclusions and Outlook

Initial research on the impact of actuator faults on large-scale wind turbine power production and structural loads is provided in this paper. The slow dynamic fault has a limited effect on the response of the turbine with an industry-standard controller. However, if the pitch actuator dynamics change too much leading to reduced performance or seized actuators, the power production will be reduced, and the structural loads will be increased significantly.

In future works, intensive research with more detailed fault cases and wind profiles can be realized to investigate the effects of the severity of each fault case. The results are used for assessing the health status of the actuator thus aiding the operations and maintenance decision making.

This research is funded by Hanoi University of Science and Technology (HUST) under project number T2022-TT-009.

References

1. M. Wilkinson, B. Hendriks, F. Spinato, K. Harman, E. Gomez, H. Bulacio, J. Roca, P. Tavner, Y. Feng, and H. Long, *Methodology and results of the reliawind reliability field study*, EWEC 2010 (Vol. **3**, pp. 1984-2004), Sheffield (2010)
2. R. Bi, K. Qian, C. Zhou, D.M. Hepburn, and J. Rong, *A survey of failures in wind turbine generator systems with focus on a wind farm in China*, Ran Bia and oth. International Journal of Smart Grid and Clean Energy (2014)

3. J Li, J Bian, Z Chuang, Y Jiang, and S Leng, *Impact of pitch actuator fault on 10-MW semi-submersible floating wind turbine*, Ocean Engineering, **254**, Elsevier (2022)
4. EE Bachynski, M Etemaddar, MI Kvittem, C Luan, and T Moan, *Dynamic analysis of floating wind turbines during pitch actuator fault, grid loss, and shutdown*, Energy Procedia **35**:210-222 (2013)
5. M Etemaddar, M Blanke, Z Gao, T Moan, *Response analysis and comparison of a spar-type floating offshore wind turbine and an onshore wind turbine under blade pitch controller faults*, Wind energy **19.1**:35-50. (2016)
6. J Jonkman, S Butterfield, W Musial, and G Scott, *Definition of a 5-MW Reference Wind Turbine for Offshore System Development*, United States. doi:10.2172/947422, https://www.osti.gov/servlets/purl/947422, (accessed 24/11/2023) (2009)
7. OpenFAST Documentation, NREL, https://openfast.readthedocs.io (accessed 24/11/2023)
8. BJ Jonkman, Turbsim user's guide v2.00.00, NREL, (2014)
9. N Abbas, D Zalkind, L Pao, and A Wright, *A reference open-source controller for fixed and floating offshore wind turbines*, Wind Energy Science Discussions 2021: 1-33. (2021)
10. SP Mulders, JW Van Wingerden, *Delft Research Controller: an open-source and community-driven wind turbine baseline controller*, Journal of Physics: Conference Series. Vol. **1037**. No. 3. IOP Publishing, (2018)

Thermal Insulation Panels with Bio-Based Adhesives

Valentina Vasilievna Strikun[1] *Mikhail Andreevich Bajandin*[1], *Vladimir Nikolaevich Ermolin*[1], and *Zoltán Pásztory*[2*]

[1]Reshetnev Siberian State University of Science and Technology, Krasnoyarsk Rabochy Av., Krasnoyarsk 660 037, Russian Federation
[2]University of Sopron, 4. Bajcsy Zs., Sopron 9400, Hungary, *E-mail address: pasztory.zoltan@uni-sopron.hu

Abstract. Environment friendly insulation panels were manufactured and tested made of pine wood fibers and glued with a bio-based adhesive, called dextran. The aim of this work was to determine relevant technical properties of panels fabricated with this new glue type. In the panels, the ratio of dry glue content was varied from 30% to 60% with 10% steps which runs served as comparison basis. The density of the insulation panels was set to 70 kg/m^3, 100 kg/m^3 and 125 kg/m^3. Beside their thermal conductivity, the compression strength, the bonding strength, and the wetting angle of the adhesive were measured. With the variation of panel densities, the thermal conductivity is in a narrow range of 0.039 to 0.042 W/mK. The resistance to compression at 10% strain was measured to be 0.3 MPa, 0.35 MPa and 0.4 MPa in the panels with 70 kg/m^3, 100 kg/m^3 and 125 kg/m^3, respectively. The wetting angle of the adhesives seemed to correlate only weakly with the bonding strength, and the glue`s wetting ability diminished with the increase of the glue content. The results seem to be competitive if compared with the traditionally used glass and rockwool, and foam insulation materials.

1 Introduction

Generally, municipal and residential buildings are responsible for a high consumption of energy [1]. Their individual energy consumption depends strongly on both the climatic conditions and the insulation characteristics of the buildings. The most effective way to improve energy efficiency of buildings is to increase the thermal insulation of their cladding. In some geographic areas, the greenhouse gas reduction and lower energy dependence - and thus, lower building operation costs - are the driving forces for stricter building codes [2]. In order to increase the energy efficiency of edifices more insulation material needs to be used resulting lower operational cost and reduced carbon-dioxide emission. Also, the production, the transportation and the elimination of surplus insulation material emit significant amount of greenhouse gases. When environment protection is a determining factor then the issue of insulation material is of high relevance. Although foam as well as rock and fiberglass based materials have a good thermal performance – usually, at low market prices, their environment impairing effects are considerable during manufacturing and at final disposal. The recycling of traditionally used foam, rock or glass wool insulation materials is still problematic, and often hazardous.

In the last decades, the research and the production of insulation panels made of natural raw materials, specifically lignin and cellulose, have come into the focus [3, 4]. Yet, the production of raw materials and of the adhesives pollute the environment. Because significant volume of adhesives is being used one of the most important goals was to lower formaldehyde emission [5]. More comprehensive works were published in the recent years concerning bio-based adhesives [6, 7,]. One of the interesting investigations deals with the activation of lignin within the constituting particles of the panels as an adhesive [8, 9]. Modified lignin is a suitable source material for adhesives with enhanced bonding properties [10, 11, 12, 13]. Norström et al. [14] have investigated a gum dispersion while Santoni and Pizzo studied vegetable protein as a possible wood adhesive [15]. A hydrophilic, eco-friendly bio-based adhesive was produced by Yuan et al. [16], and a mimosa tannin-based adhesive by Zhank at al. [17]. Although, these research endeavors have been conducted for longer than a decade no break-through could be realized on the market.

Organic fluids have been studied as a possible alternative for adhesives, having obtained from cultivated molasses. They contain the macro-molecular substances eventually yielding dextran. Dextran is a polymer – with a mass more than 15 kilodaltons - having highly branching molecules which explains its gluing ability. The number and the positions of the branch connections and the length of the side chains significantly determine the adhesive`s strength. Besides the strong adhesive capacity, it is non-toxic, thus completely safe for humans, and entirely biodegradable. Earlier studies on particle boards using organic liquids as a binder indicate the potential use of this type of glue in chipboard production [18]. Hosseinpourpia [19] reached at similar conclusions when studying fiberboards manufactured with the adhesive dextran. Presently, the use of carbohydrates-based adhesives (modified starch, dextrin) in the production of particle boards is being widely studied and seem to harbor promising prospects [20].

The overall objectives of the present research are hence 1) to investigate the wetting angle in dependence of the dry content of dextran, 2) to determine the correlation between dry content of the bio-based dextran adhesives and their bonding strength, 3) and to measure the resistance to compressive forces at 10% strain with different densities of the panels, 4) to determine the thermal conductivity of the insulation panels as manufactured with different densities.

2 Materials and methods

2.1 Sample preparation

Dark brown, organic aqueous solution of the glue was used for this investigation. In the initial stage, the physico-chemical characteristics of the adhesive's interaction with wood fibers have been studied. Subsequently, the efforts were concentrated on the manner of fabrication of the thermal insulating boards to arrive at the necessary technical characteristics. The dextran was produced by cultivating the bacterium Leuconostoc mesenteroides. Eventually, the cultivation stock was dried and ground to a fine powder. From this powder, aqueous solutions were prepared containing 30%, 40%, 50%, 60% dry glue powder by weight ratio. This liquid was used as adhesive and added to the fiber in 30% wight related to the dry fiber and were mixed. The matt were formed in a trey having a rectangular shape panel with constant thickness and parallel surfaces. In this manner, a series of insulation panels with differing densities were produced. Eight samples were manufactured for each type of dry glue content The thermal insulation panels were produced from fibers of Siberian pine (Pinus sibirica) with a moisture content of 2-4%, donated by the MDF Factory in Krasnoyarsk region of Russia. The glue was sprinkled onto the rotating fiber mass by a pneumatic mixer. The mat was formed manually, and by using a laboratory press panels with a surface size of 400 mm by 400 mm and a thickness of 25 mm were produced (Fig. 1). The mat was subjected to a pressure of 0.2 MPa for five minutes without heating. Because of the cold pressing, no vapor overpressure had built up inside the panel, and thus the press could be opened without pressure reduction steps. Thereafter the panels were placed into a drying chamber with air circulation and exposed to 105°C for 24 hours.

Fig. 1. Insulation panel with dextran adhesive

2.2 Experimental setup for wetting angle measurement in dependence of dry glue content

Wetting behavior of the glue can significantly influence the strength of the bond. For this reason, the wetting angle of glue was studied with different dry contents. For the measurements, a glue was dropped with a glass rod onto the planed surface of a solid pine board near its edge. A reconstructed MIS-11 microscope was used to determine the wetting angle. The diameter and the height of the glue drop were determined with an ocular micrometer. The angle of the wetting edge was calculated by the formula 1:

$$\cos X = (r2 - h2) / (r2 + h2) \quad (1)$$

where, X is the wetting angle in degree, r is the radius and h the height of the droplet, both in μm. 50 tests were carried out for every dry content ratio of the glue, and the usual statistics calculated.

2.3 Compression strength

Compression strength tests were carried out on the panels as an indication of the adhesive's bonding strength. To this procedure, the 25 mm thick samples were cut to a surface size of 100 mm x 100 mm. 20 repetitions were prepared for each glue content variation. The results were calculated as suggested by the standard GOST 17177-94 [21] according to the equation 2:

$$\sigma_{10} = P/lb \quad (2)$$

where σ10 is the compression stress level in MPa at the linear strain of 10%, P is the load in Newton; l sample length and b sample width, both in mm.

2.4 Measurement of the panel´s stress level at 10% strain in dependence of dry glue content

Panels have been tested to study the effect of the dry glue content on the stress level at 10% strain. To this goal, eight pieces of five-layer plywood samples were prepared according to the standard EN 314-1:2004 [22] for all four dry content ratios of the glue. Prior to testing, the samples have been stored in the climate chamber until they have reached the required moisture content of 12%. Altogether, thirty-two panel samples were tested. Compression strength tests were carried out by using a universal testing machine USN-30 with a measurement accuracy of 0.01 Newton, and 0.01 mm.

2.5 Measurement of thermal conductivity of the panels

For the determination of the thermal conductivity coefficient, the 25 mm thick panels samples were cut to the dimensions of the measuring device's working surface of 150 mm by 150 mm. The tests were carried out according to the requirements of the standard GOST 7076-99 [23] at steady state condition and an ambient temperature of 22°C and 65% relative humidity, respectively. 20 samples were cut out from the pressed panels for each variation of density and dry content of the glue. The results were calculated as the average of the 20 individual results.

3 Results and discussion

3.1 Wetting angle measurement in dependence of dry glue content

In Table1, the data for the equilibrium wetting angles represent mean values of 50 measurements. The values in parentheses (standard deviations) indicate that the angles significantly differ from each other, that the dry glue content alters indeed the interaction with the wood fibres.

Table 1. Equilibrium wetting angle of bio-based adhesive with different dry contents

Type of substrate	Dry glue content, [%]	Radius of the droplet, [μm]	Height of droplet, [μm]	Equilibrium wetting angle, [°]
Coniferous wood (Siberian pine fibres)	30	158 (±2.162)	75 (±1.969)	50.785 (±1.266)
	40	175 (±1.871)	91 (±0.700)	52.345 (±0.531)
	50	211 (±2.462)	114 (±1.256)	55.089 (±0.772)
	60	241 (±2.370)	131 (±2.118)	57.054 (±0.817)

As the dry content of the glue increases the wetting angle rises too which means the wetting ability is decreasing. According to Fig. 2, the interdependence seems to be linear. The wetting ability of the glue with 30% dry content is the highest as indicated by the wetting angle (when 180°stands for water-repellence and 0°for full solubility). Yet, when the dry content is doubled from 30% to 60% the wetting angle increases only slightly, by about 6 + degrees. Thus, the increasing dry glue content makes the panels only marginally more hydrophilic.

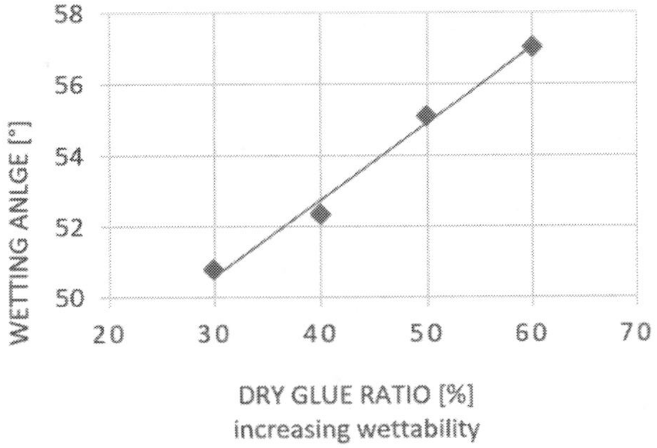

Fig 2. Wetting angle [deg] in function of dry glue content [%]

3.2 The effect of dry content of glue on the adhesive bond strength

The data in Fig. 3 shows the average data of the 20 samples from the 100 kg/m³ density panel according to the dry content of the adhesive.

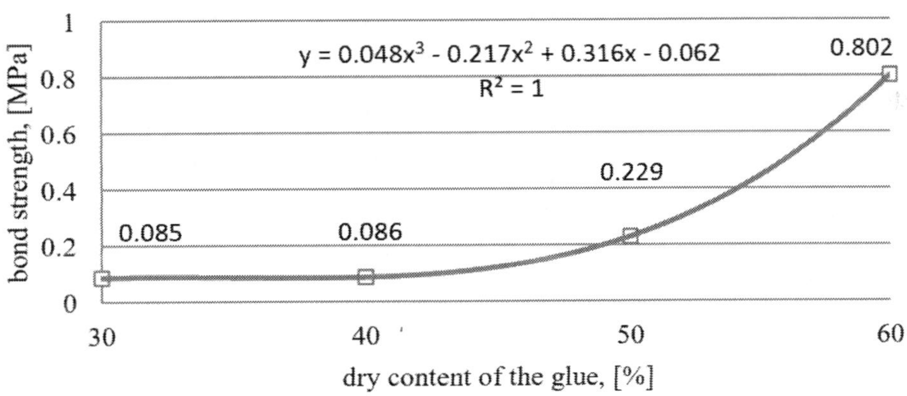

Fig. 3. Influence of the dry content of glue on tensile

The fig. 3 shows that the panel strength starts to increase as from 40% dry glue content. The steepness of the function line increases with higher dry glue content. The higher polymer content of the glue shows the increased adhesion ability of the solution. It should be noted that the high content of dry component increases not only its adhesive ability, but also simultaneously increases the viscosity of the glue, which made the process of wetting of wood fibers more difficult. Wood is a porous material full of capillaries, and when glue can intrude into these voids, the bonding strength increases. The amount of moisture absorbed by the wood depends on both the initial water content of the glue and the moisture content of the fibers. Accordingly, with low dry glue content more water will be absorbed by the panel.

That increases the concentration of dry substances on a surface of the fiber particles and contributes to reach strong bonding connection. Despite of this theory the results showed that the glue with more than 50% dry content shows a significant bonding strength.

3.3 Measurement of the resistance against compressibility

The mechanical compressibility of the panels is an important information for the builder. The resistance to mechanical compressibility of the panels were measured at 10% of the strain as derived from the data in Table 2.

Table 2. Compression strength at 10% compression

	Panel density		
	70 kg/m³	100 kg/m³	125 kg/m³
Stress at 10% strain [MPa]	0.030 (±0.002)	0.035 (±0.004)	0.040 (±0.003)

3.4 Thermal conductivity

Intuitively, a strong interdependence between thermal conductivity and resistance against mechanical compressibility of the panels is presumed. According to our results, however, the thermal conductivity of the panels seemed to increase significantly while the stress at 10% strain level rose only marginally from one density step to the next. When the panel density changed from 70 kg/m³ to 125 kg/m³ the thermal conductivity was enhanced by merely 0.003 W/mK (Table 3). In contrast, the compression stress at 10% strain increased by about one third while the panel density nearly doubled from 70 kg/m³ to 125 kg/m³. It is a significant increase indeed as is shown in Fig. 4.

Table 3. Properties of thermal-insulating panels using 30% dry content of glue

	Panels density		
	70 kg/m³	100 kg/m³	125 kg/m³
Thermal conductivity [W/mK]	0.039 (±0.003)	0.040 (±0.004)	0.042 (±0.003)

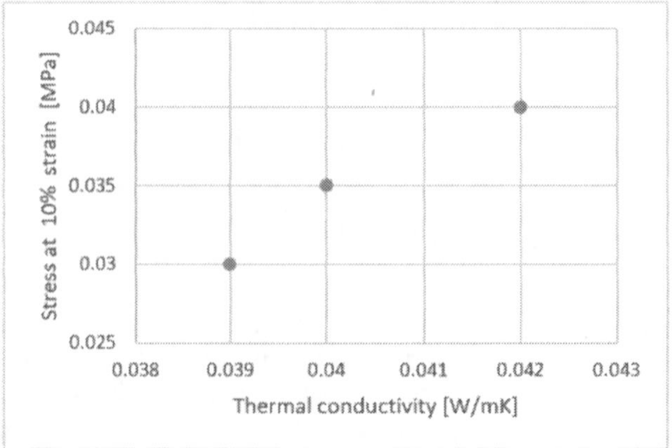

Fig. 4. Interdependence between thermal conductivity and 10% mechanical compressibility

4 Conclusion

The overall goal of this study was to investigate the different parameters of thermal insulation panels made of wood fiber and dextran as a bio-based glue, at varying dry glue contents. The results showed the following conclusions:

1) The dry content of bio-based adhesive significantly affects the bonding strength in the insulation panels. Generally, the higher the dry content the higher is the panel`s bonding strength which seems non-linearly increase with the change of dry glue content from 30% to 60%.
2) By augmenting the dry glue content in the panel, the wetting angle rises only slightly. The angle increase is small, about 6°, while the glue dry content doubles from 30% to 60%.
3) Also, the thermal conductivity has increased by a small value while the panel density changed from 70 kg/m^3 to 125 kg/m^3.
4) Resistance against mechanical compressibility increased moderately with the increased density of the insulation panels.

Dextran is a suitable adhesive for producing sustainable fiber-based insulation panels, with higher dry content of at least 50% or rather 60%. The wetting angle is less important than it was supposed. The higher the dry content, the higher the wetting angle, although the bonding strength will be higher also. The results indicate that this bio-based adhesive is suitable for producing environment friendlier insulation material than most of them on the market. It has more advantageous: harmless the environment during manufacturing process, provides healthier living conditions because no VOC emission, and after life cycle can be reuse or recycle without any environment damage. The expected life cycle of this new insulation material glued with bio-based adhesive can be the same than that of the normal glass wool or polystyrene insulations. Beside these facts the thermal conductivity is almost the same than that of the commercialized insulation materials and the 10% compression strength is high enough for using under floor. In the future studies are planned to determine the optimal size of wood fiber used as a filler. For the practical use of this thermal insulation material, there must be research on its fire resistance and durability. Secondly it would be important to estimate the production cost and the carbon dioxide equivalence and compere to the competing fiber-based insulation already in the market.

Acknowledgement

This article was made in frame of the project TKP2021-NKTA-43 which has been implemented with the support provided by the Ministry of Innovation and Technology of Hungary (successor: Ministry of Culture and Innovation of Hungary) from the National Research, Development and Innovation Fund, financed under the TKP2021-NKTA funding scheme.

References

1. Schiavoni, S., D'Alessandro, F., Bianchi, F., Asdrubali, F. Insulation materials for the building sector: A review and comparative analysis. *Renew. Sust. Energ*, 62, 988-1011. (2016)
2. DIRECTIVE 2010/31/EU of the European Parliament and of the Council of 19 May 2010 on the energy performance of buildings.

3. Mimini, V., Kabrelian, V., Fackler, K., Hettegger, H., Potthast, A., Rosenau, T. Lignin-based foams as insulation materials: a review. *Holzforschung,* 73, 117-130. (2019) https://doi.org/10.1515/hf-2018-0111
4. Jones, D. COST FP1303 "performance of bio-based building materials" FOREWORD. *Wood Material Science & Engineering,* **14**, 1-2. Special Issue (2019) https://doi:10.1080/17480272.2018.1528569
5. Mantanis, G.I., Athanassiadou, E.T., Barbu, M.C., Wijnendaele, K. Adhesive systems used in the European particleboard, MDF and OSB industries. Wood Material Science & Engineering, 13, 104-116. (2018) https://doi:10.1080/17480272.2017.1396622
6. Zhongqi, He. *Bio-based Wood Adhesives, Preparation, Characterization, and Testing*. Taylor &Francis Group; (2017)
7. Gadhave, R.V., Mahanwar, P.A., Gadekar, P.T. Starch-Based Adhesives for Wood/Wood Composite Bonding: Review. *Open Journal of Polymer Chemistry*, **7**, 19-32. (2017) https://doi.org/10.4236/ojpchem.2017.72002
8. Domínguez-Roblesa, J., Tarré, Q., Delgado-Aguilar, M., Rodríguez, A., Espinach, F. X., Mutjé, P. Approaching a new generation of fiberboards taking advantage of selflignin as green adhesive. *International Journal of Biological Macromolecules*, **108**, 927–935. (2018) https://doi.org/10.1016/j.ijbiomac.2017.11.005
9. Ghaffar, H.S., Fan, M. Lignin in straw and its applications as an adhesive. *International Journal of Adhesion & Adhesives*, **48**, 92–101. (2014) http://dx.doi.org/10.1016/j.ijadhadh.2013.09.001
10. Ibrahim, V., Mamo, G., Gustafsson, P.J., Hatti-Kaul, R. Production and properties of adhesives formulated from laccase modified Kraft lignin. *Industrial Crops and Products*, **45**, 343– 348. (2013) http://dx.doi.org/10.1016/j.indcrop.2012.12.051
11. Ji, X., Guo, M. Preparation and properties of a chitosan-lignin wood adhesive. *International Journal of Adhesion and Adhesives*, **82**, 8–13. (2018) https://doi.org/10.1016/j.ijadhadh.2017.12.005
12. Jin, Y., Cheng, X., Zheng, Z. Preparation and characterization of phenol–formaldehyde adhesives modified with enzymatic hydrolysis lignin. *Bioresource Technology*, **101**, 2046–2048. (2010) https://doi.org/10.1016/j.biortech.2009.09.085
13. Pradyawong, S., Qi, G., Li, N., Sun, X. S., Wang, D. Adhesion properties of soy protein adhesives enhanced by biomass lignin. *International Journal of Adhesion and Adhesives*, **75**: 66–73. (2017) http://dx.doi.org/10.1016/j.ijadhadh.2017.02.017
14. Norström, E., Fogelström, L., Nordqvist, P., Khabbaz, F., Malmström, E. Gum dispersions as environmentally friendly wood adhesives. *Industrial Crops and Products*, **52**, 736– 744. (2014).http://dx.doi.org/10.1016/j.indcrop.2013.12.001
15. Santoni, I., Pizzo, B. Evaluation of alternative vegetable proteins as wood adhesives. *Industrial Crops and Products*, **45**, 148– 154. (2013) http://dx.doi.org/10.1016/j.indcrop.2012.12.016
16. Yuan, C., Chen, M., Luo, J., Li, X., Gao, Q., Li, J. A novel water-based process produces eco-friendly bio-adhesive made from green cross-linked soybean soluble polysaccharide and soy protein. *Carbohydrate Polymers*, **169**, 417–425. (2017) http://dx.doi.org/10.1016/j.carbpol.2017.04.058
17. Zhang, J., Xiong, L., Zhou, X., Du, G. Development of Mimosa Tannin-Based Adhesive Cross-Linked By Furfuryl Alcohol-Formaldehyde and Epoxy Resins. *Wood Research*, **64**, 847-858. (2019)

18. Pashkin, S.V., Ivanova, M. A., Shchekolov, V. M. Innovative technologies for producing environmentally friendly glued materials based on water-soluble bio-glue from low-grade wood. *Lesnoj vestnik (Forest Bulletin)*, **8**, 59-63. (2012)
19. Hosseinpourpia, R., Adamopoulos, S., Mai, C., Taghiyari, H. R. Properties of Medium-Density Fibreboards Bonded With Dextrin-Based Wood Adhesive, *Wood Research*, **64**, 185-194. (2019)
20. Bajandin, M.A. Chipboard on a binder modified dextran. / M.A. Bajandin // *Aktual'nye napravleniy anauchny hissledovanij XXI veka: teoriyaipraktika* (Current research areas of the XXI century: theory and practice). 13-1, 355-359. (2015)
21. GOST 17177-94 (1996): Материалы и изделиястроительныетеплоизоляционныеМетодыиспытаний (Thermal insulating materials and products for building application. Test methods)
22. EN 314-1:2004 Plywood. Bonding quality Test methods, the standard is in harmony with GOST 15613.1-84 (1984): ГОСТ 15613.1-84 Древесинаклееная массивная. Методы определения предела прочности клеевого соединения при скалывании вдоль волокон, (Gluedmassivewood. Methods for determination of glued joint of ultimate strength while shearing along grain)
23. GOST 7076-99 (1999): ГОСТ 7076-99. Материалы и изделия строительные. Метод определения теплопроводности и термического сопротивления при стационарном тепловом режиме, (Buildingmaterialsandproducts. Method of determination of steady-state thermal conductivity and thermal resistance)

Wood Research at the University of Sopron – Physical-mechanical Properties

Szabolcs Komán [1], *Róbert Németh* [1], *James Kudjo Govina* [1], *Ágnes Vörös* [1*], *Dénes Ákos Horváth* [1], and *Sándor Fehér* [1]

[1]University of Sopron, Faculty of Wood Engineering and Creative Industries, 9400 Sopron, Hungary

Abstract. This study series shows research fields and results in wood science of the University of Sopron, Hungary. In this paper, we review several studies regarding the physical-mechanical properties of wood, including some properties of Turkey oak grown in different stands, showing that it should be planted among other species in same stand to achieve optimal wood quality. Research in the field of the firewood plantations has shown that it is necessary to choose the right species of poplars, in a given area. Paulownia and some invasive species grow rapidly, so they have great potential as raw materials and there are significant differences between paulownia varieties. Based on their densities, box elder, tree of heaven and green ash can be substituted for the noble species. Basswood can be well impregnated with paraffin, making it a good choice for pencils. Low-quality noble oak logs have the potential to be used in structural elements as lamellae, improving the yield and decreasing the loss in the wood industry. Black locust, sessile oak, beech, poplar, larch and spruce were subjected to different mechanical test methods at different moisture contents. Their properties decreased with increasing moisture content between 12% moisture content and FSP for all six wood species, but the ratio of changes varied, depending both on the measurement method and on the species.

1 Introduction

The University of Sopron in Hungary deals extensively with the proper management of trees and wood from both scientific and practical perspectives. The Faculty of Forestry is the only institution in Hungary that trains, among others, forest engineers and nature conservation engineers, while the Faculty of Wood Engineering and Creative Industries trains, among others, engineers for the wood industry, specialists in the creative industry, and product designers. The former Institute of Wood Science, now part of the Institute of Basic Sciences, deals with the properties and modification possibilities of wood. The following scientific publications are good examples: [1–17]. This study presents the activities and some of our important results we have carried out in recent years to present research regarding the wood industry. We believe that the work we do is globally important. This article covers density, modulus of rupture and modulus of elasticity results for Turkey oak grown in both mixed-species and pure-species stands; energetical utilization research of plantation wood;

* Corresponding author: voros.agnes@phd.uni-sopron.hu

characterization of paulownia wood and some invasive wood species; impregnation of basswood with paraffin; potential uses of low-quality oak; and compares mechanical properties of wood at different moisture contents.

1.1 The influence of stand composition on the selected properties of Turkey oak (*Quercus cerris* L.)

Global forests are under pressure to produce adequate timber to meet the increasing demand for wood and wood products. Matching forest resources to consumer demand has always been complex and of a great concern. This situation gives relevance the concept of sustainability, which can only be achieved through a robust system of forest management. Studies have predicted that global timber consumption will increase by at least 40% between 2005 and 2030 [18–20]. Meanwhile deforestation and forest degradation are still global threats and contribute to about 15% of global warming. More frequent and severe droughts have been predicted for our future climate with expected repercussion on the resilience of forest ecosystems [21–23]. Timber trade regulations and loud environmental concerns are likely to shift the supply of wood from natural forests to plantation timber. Turkey oak species are categorized as drought tolerant and are under-utilized in Hungary. Potentially, Turkey oak can be a promising species for the Hungarian industry because of the predicted climate change. However, the effect of tree species composition on wood quality is limited for Turkey oak. Such information is needed for effective decision making in forest operations. Wood density is an important trait that indicates several other wood properties, as well as the quality of the given species.

The demand for wood as a raw material is increasing and this trend also reflected in research directions. There is a growing demand for rapidly growing tree species that yield large amounts of timber quickly, and their growth characteristics are just as important as the properties of the wood itself. New varieties may also be important within the energy sector as their yields per hectare and their heating value together determine their energy yield. Furthermore, the broader utilization of existing resources is facilitated by the application of existing species in new areas.

1.2 Utilization of wood in the energy field

Biologically renewable energy sources are increasingly significant as a result of ongoing rearrangements within the energy sector. Forest plantations, including those recognized by the Kyoto Protocol as resources to reduce greenhouse gas emissions [24], belong to this category. Rapidly growing tree species such as poplar, willow, and black locust are particularly important for firewood plantations because of their high biomass production and excellent sprouting ability. These short rotation plantations can generate significant quantities of wood as a renewable energy source in a relatively short time period. To make full use of suitable land for establishing plantations, it is essential to select the most appropriate tree species or varieties based on local conditions. However, several factors must be considered when making these selections.

1.3 Characteristics of paulownia (*Paulownia* spp.) wood

There is a growing demand for fast-growing tree species to fill the gap in the wood trade between supply and demand. Prominent representatives among these are several varieties of Paulownia trees, which belong to the fastest growing tree species in the world, but also have one of the lowest densities, with a significant portion of pith. Interest in Paulownia trees has significantly increased across Europe in recent decades. However, the available scientific publications do not provide a complete view of the wood characteristics of this tree species. For different cultivated varieties, reference is primarily made to the characteristics of the basic species. Among the characteristics of wood, density is the most crucial for determining its usability. Knowledge of density predicts its strength properties [25].

1.4 Characteristics of invasive wood species

The spread and habitat-transforming effects of invasive plant species cause significant environmental problems worldwide [26]. Efforts to combat invasive tree species are being made globally, but necessary basic information for their wood utilization is less known to the processing industry. The most critical characteristic for the usability of wood is its density, from which we can make inferences to its strength properties. The literature on invasiveness has rapidly expanded over the past decades, but it does not typically delve into the wood properties of individual species. Knowledge of of densities is one of the most crucial attributes for the industrial use of tree species for wood, paper, and energy. Density is closely linked to the quality parameters of end-use, such as cellulose yield and the strength of building materials [27]. To position the densities of the three invasive tree species in relation to other tree species, it is advisable to compare the results within the same genus but of greater significance tree species.

1.5 Impregnation of basswood (*Tilia argentea*) for use in pencil manufacturing

There are special requirements concerning raw materials for pencils. They include a uniform structure, relatively small differences between earlywood and latewood density, straight grain and a lack of defects, especially knots. As a consequence, there are a relatively small number of species suitable for pencil manufacturing. The technology also requires relatively large cross-sections and pith-free lumber for producing the lamellae used for pencil manufacturing. Therefore, the raw material base available for pencil manufacturing is restricted, and there is a competition amongst pencil producers for raw materials. Basswood is homogenous and smooth in texture, making its mechanical processing and surface treatment trouble-free. Timing is essential when applying paraffin treatment to basswood for pencil manufacturing.

1.6 Visual assessment of boards made from low-quality noble oak (*Quercus* spp.) logs and the use of noble oak lamellae produced from low-quality timber

In this study, we tested the quality of timber derived from low-quality oak logs. Our main objective was to determine the amount of defect-free lamellae that can be obtained from logs that do not reach category II of the current Hungarian standard (MSZ 45 [28]).

Currently, lower quality hardwood lamellae (including knots, sapwood, etc.) are rarely used for structural purposes. Their most common use is for energy purposes, although a reassessment of their use, based on habits and traditional practices, could allow their higher value recovery, improving plant performance and profits, partly reducing the shortage of timber and contributing to more environmentally friendly operations.

 https://doi.org/10.1051/e3sconf/202451404002

1.7 Comparison of selected mechanical properties of six wood species at different moisture contents

The properties of wood are the result of several factors. The structure of annual rings and the resulting inhomogeneity, the porosity, the moisture content (*MC*) and many other factors influence wood properties.

Water can be found in wood in three states: liquid water (free water) and water vapor in cell lumina; water bound in cell walls (bound water); and crystallized water linked to the chemical components of wood. The fiber-saturation point (*FSP*) is the *MC*, when all the cavities of the cell wall are saturated with water, but no free water is in the cell lumen (Molnár 2004). *FSP* is at about 30% *MC*, depends on the wood species. The mechanical properties of wood change a lot below the *FSP*. The aim of this study was to determine the differences of tensile strength, bending strength, compressive strength and Brinell hardness between 12% *MC* and *FSP*.

2 Materials and methods

2.1 The influence of stand composition on the selected properties of Turkey oak

A total of 12 Turkey oak (*Quercus cerris* L.) trees were harvested from Vas County in Hungary. 7 trees from pure-species stand and 5 trees from a mixed species stand. The selected trees were at least 67 years old. Samples prepared for testing, according to standards, were taken from breast-height level with consideration for both sapwood and heartwood to cover possible variability within trees. The moisture content of the test samples was 12% after a conditioning at normal climate (20 °C and 65% RH).

2.2 Utilization of wood in the energy field

This study included three-year-old clones of three different hybrid poplar (*Populus*) varieties ('Erti', 'I-214', 'Kopeczky') from two distinct growing areas (Borjád and Sárvár). After assessing yields and heating values, the expected heating values per hectare were determined.

2.3 Impregnation of basswood for use in pencil industry

The experimental basswood samples were impregnated with paraffin emulsion at a net *MC* of 7%, using three different methods. The applied vacuum and pressure parameters were the same in all three cases.

2.4 Producing high added value structural timber from lamellae waste

100 pieces of small oak specimens (*Quercus* spp.; 22 × 50 × 424 mm) remaining as by-products of an industrial production were subjected to both non-destructive and bending tests. The analysis of the non-destructive tests showed that the method used does not provide reliable results due to the small specimen sizes and is therefore presented for information only. The requirements of standard EN 408: 2010 + A1 [29] were followed for bending tests because it deals both with structural solid and glued-laminated timber and the end-use of small-size lamellae may be in load-bearing structures. The bending tests were performed on an Instron 4208 (Instron Corporation, USA) universal material testing machine.

2.5 Comparison of selected mechanical properties of six wood species at different moisture contents

For the tests, the most commonly used Hungarian hardwood and softwood species were selected: black locust (*Robinia pseudocacia*), sessile oak (*Quercus petraea*), beech (*Fagus sylvatica*), poplar (*Populus* spp.), larch (*Larix decidua*) and spruce (*Picea abies*). The porosity and density of the samples were divided into three categories:

- high density (>700 kg/m^3): black locust, beech
- medium density (550-700 kg/m^3): oak, larch
- low density (<550 kg/m^3): poplar, spruce.

Tensile strength tests were done according to standard ISO13061-06 [30]. Bending tests followed the specifications of ISO13061-03 [31]. Compression tests were carried out according to ISO13061-17 [32] standard. Brinell-Mörath hardness tests were done as it is shown in the standard MSZ 6786-11 [33].

3 Results and discussion

3.1 The influence of stand composition on the selected properties of Turkey oak

ANOVA analysis showed that there was a statistically significant difference between turkey oaks grown in mixed and pure species stands for the three wood properties (basic density; modulus of rupture; modulus of elasticity). The outputs in Figure 1 compare the effect of stand composition on basic density, modulus of rupture and modulus of elasticity, respectively.

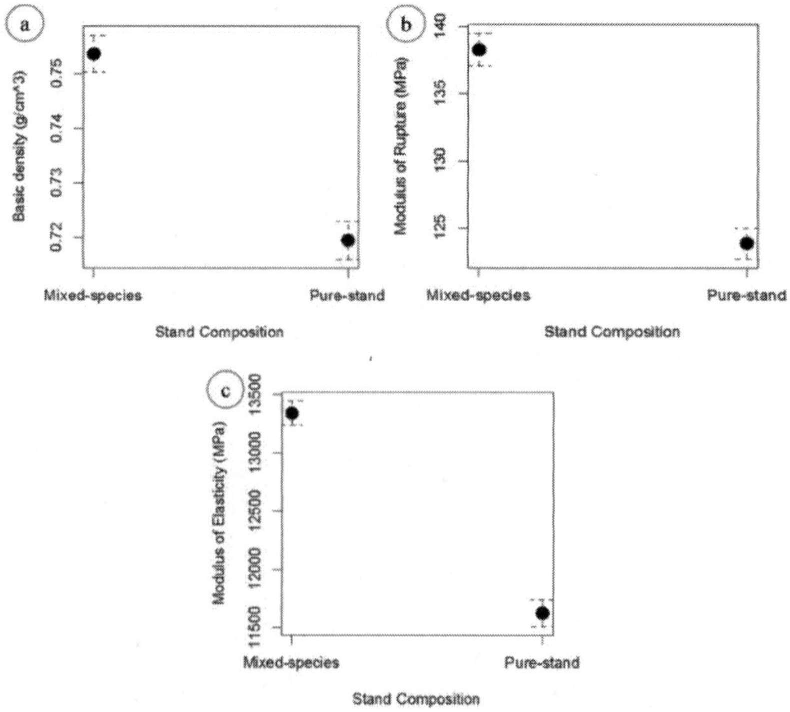

Fig. 1. Basic density (a), modulus of rupture (b) and modulus of elasticity (c) for Turkey oak from mixed- and pure-species stands from Vas County, Hungary.

This study indicates that plantation wood from Turkey oak exhibit traits suitable construction purposes. However, to guarantee superior qualities, Turkey oak should be planted among other species in same stand.

3.2 Utilization of wood in the energy field

Based on the results of the research, there is no significant difference in the lower heating values (LHV) of different varieties when projected on weight, and the quality of the growing site does not have a substantial impact within the same variety (Figure 2). However, from the perspective of yields, the quality of the growing site is important. Energy yield is determined by these two components, and the result can be greatly modified by the proper variety selection in a given area, where nearly twice the yield can be achieved.

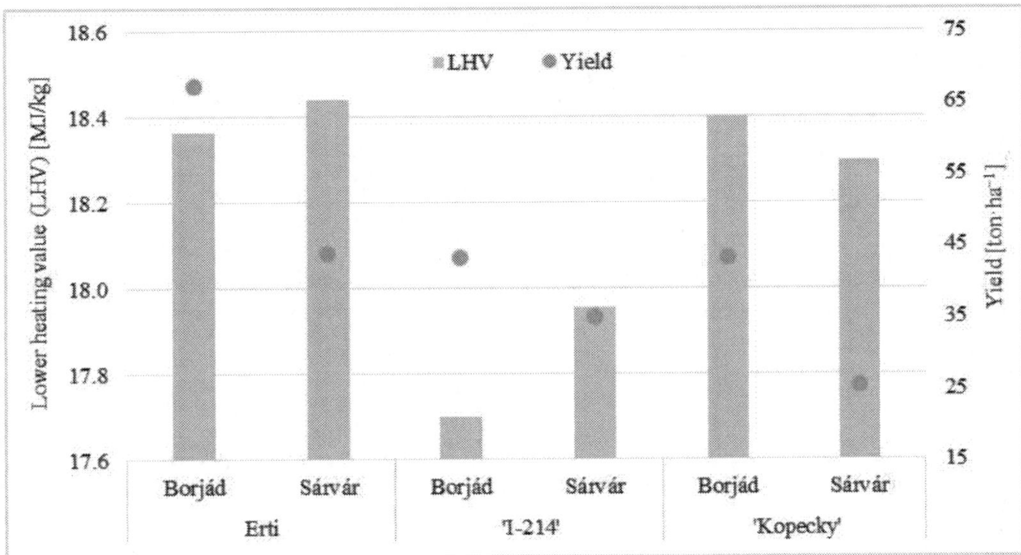

Fig. 2. The heating value and yield of hybrid poplars.

3.3 Characteristics of paulownia wood

The base species *Paulownia tomentosa*, as well as the *Paulownia* Clone in vitro 112 and *Paulownia tomentosa* Robust4 variants studied so far, all exhibit very low densities (Figure 3). However, there can be significant differences between individual varieties, is important for industrial utilization. In the low density of the wood, the strength properties of paulownia are also moderate.

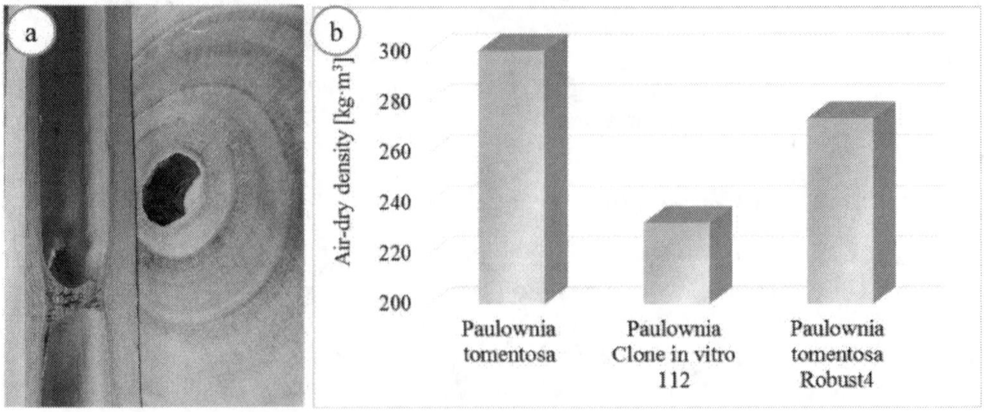

Fig. 3. The wood of Paulownia (a) and the air-dry densities of paulownia species (b).

3.4 Characteristics of invasive wood species

To position the densities of these three invasive tree species in relation to other tree species, it is advisable to compare the results within the same genus but of species which have greater significance in the industry and wood science. The densities of these three invasive tree species are well separated from each other (Figure 4). The lowest value is the box elder (*Acer negundo*), followed by the two ring-porous species, the tree of heaven (*Ailanthus altissima*), and the green ash (*Fraxinus pennsylvanica*) in sequence. The air-dry density of the green ash is similar to that of the common ash, but the box elder's density may be significantly lower than that of sycamore.

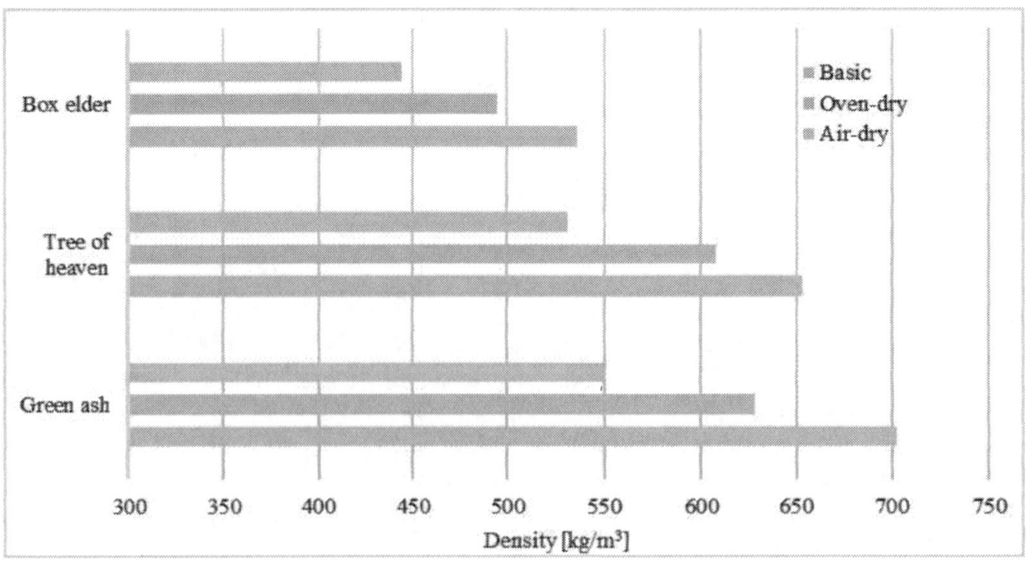

Fig. 4. Different density values of the examined invasive tree species.

3.5 Impregnation of basswood for use in pencil manufacturing

The experimental results suggest that the lower the wood density, the greater the amount of paraffin is taken up (Figure 5). At around 400 kg/m^3 wood density, the wood can take up nearly twice its original weight in paraffin, while at around 500 kg/m^3 density, it can only uptake less than half. There is no significant difference in impregnation quality among the three different impregnation methods applied, described in Figure 5.

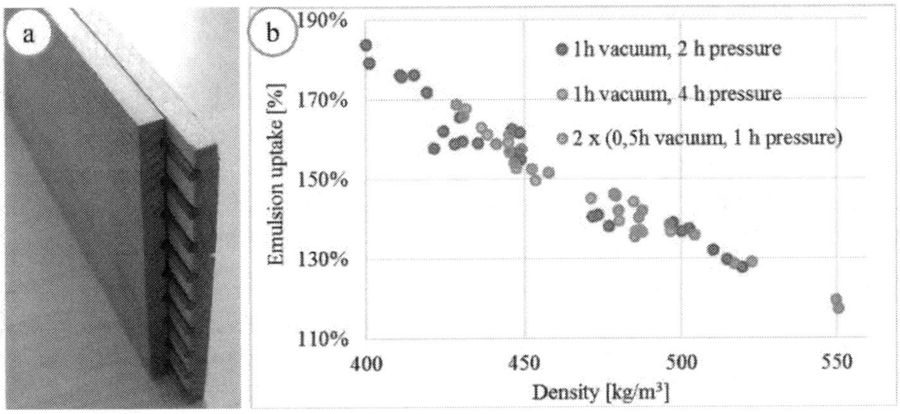

Fig. 5. Impregnated pencil raw material (a) and the relationship between the density and emulsion uptake (b).

3.6 Visual assessment of boards made from low-quality oak logs

Out of 50 boards with a surface area of approximately 60 m^2, 11 m^2 of lamella surface could be extracted, which represents an extraction rate of 18%. The survey was carried out with the lamellae widths and lengths in use today, with the largest proportion of lamellae being the middle size, 50 mm wide. This is definitely advantageous from a value-added point of view. Three quarters of the assortment was between 0.25 and 0.5 metres long, i.e. belonging to the short lamella section. All these results were due to the presence of a significant number of wood defects which limited the extraction of first grade lamellae: mainly sound and dead knots, slope of grain, sapwood, fissures and biotic damages. Given that the subsequent use of these small-sized lamellae seems possible mainly in glued-laminated wood supports, it is likely that lower quality lamellae (contain small knots, and sapwood, etc.) could be used as well, which would highly increase the yield. This requires further research, analysis and related mechanical testing.

3.7 Production possibilities of high added value structural timber from lamellae waste

The average lamellae density (712.6 ± 72.5 kg/m^3) was within the published values, despite the wood defects, but a density test alone cannot be used for classification. The relationships between density and both static and dynamic modulus of elasticity of the specimens were good (R^2 = 56.4% and 51.3%, respectively). There was a good linear relationship between static and dynamic modulus of elasticity, too (R^2 = 57.5%). The best correlation was between the deflection measured in the elastic range and bending strength (R^2 = 62.8%). These lamellae may be used mainly to produce glued-laminated timber, which would allow part of the wood nowadays used as firewood to be sold as a very high value-added product. This idea is supported by the classification obtained according to standard for structural timber EN 338 [34]: this sample set belongs to strength class D35 and it is particularly suitable for further processing. Visual grading criteria have also been defined to visually pre-grade our sample set currently being taken out of industrial production. By doing this, 20% of the specimens were declared as unusable and the classification of the remaining sample set is significantly improved as strength class D45. Comparing our average results with those of the clear samples, they are very similar, meaning that there is a large amount of additional wood available for further industrial use. Taking into account our results, it is likely that a significant amount of the industrial timber residues could be used for lamellae, including for glued laminated structural applications. Of course, several other products could also be made by finger jointing, panelling, etc. In this way, a higher proportion of timber could be utilized in the wood industry, which creates the potential for more profitable management and better compliance with societal requirements.

3.8 Comparison of selected mechanical properties of six wood species at different moisture contents

Tensile strengths at 12% *MC* are well above the average values of the literature, except for oak and poplar. As example, the tensile strength can be as high as 245 MPa for spruce and 180 MPa for beech and oak. The loss of strength due to the increasing *MC* is 56.8 MPa (38%) for spruce, 18.6 MPa (16%) for larch, 55.5 MPa (26%) for black locust, 35.1 MPa (31%) for oak, 50.6 MPa (28%) for beech and 8.3 MPa (7%) for poplar.

The bending strength results of spruce, black locust and poplar exceed the mean values, while larch is much lower compared to the results of Molnár [35]. With increasing moisture content between 12% and *FSP*, bending strength decreases by 41.4 MPa (45%) for spruce, 51 MPa (52%) for larch, 56.1 MPa (39%) for black locust, 38.8 (36%) for oak, 62.8 MPa (55%) for beech and 42 MPa (49%) for poplar.

The compression strength showed a significant average reduction of 60% with the change in *MC* between 12% and *FSP*. The compression strength of spruce and poplar samples exceed the average values, while the remaining wood species remain below that. The minimum values given by Molnár [35], were always achieved.

The hardness of the wood species was tested in all three anatomical directions. The reduction of hardness with increasing *MC* is shown in Table 1.

Table 1. Changes of Brinell-Mörath hardness in different anatomical directions between 12% moisture content and fiber saturation point of different wood species.

Direction	*Black locust*	*Oak*	*Beech*	*Poplar*	*Larch*	*Spruce*
Longitudinal	41%	48%	11%	28%	33%	19%
Radial	4%	35%	7%	41%	30%	29%
Tangential	8%	43%	3%	37%	17%	20%

4 Conclusions

This paper summarizes the results of several studies, including density, modulus of rupture and modulus of elasticity for Turkey oak grown in both mixed-species and pure-species stands; energetical utilization research of plantation wood; characterization of paulownia wood and some invasive wood species; basswood impregnation possibilities with paraffin; potential uses of low-quality oak; mechanical properties compared at different moisture contents. Our purpose was to show research activities at the University of Sopron, regarding the physical-mechanical wood properties.

Turkey oak, as a wood species possibly spread in the near future in Hungary, has should be planted among other species in same stand to achieve optimal wood quality. This was evidenced by all examined properties: density, modulus of rupture and modulus of elasticity.

From the perspective of yields, the quality of the growing site does play a significant influencing role in energy-wood plantations. Nearly twice the yield can be achieved by the proper variety selection in a given area using poplar clones (*Populus*).

Paulownia belongs to the fastest-growing tree species, thus, it have a great industrial potential despite that it has a very low density, with a significant pith portion. After the examination of densities, it can be stated that there can be significant differences between individual varieties.

The study of invasives is important because their rapid spread could provide industry with a significant amount of timber in the future. By determining the density, important strength and other properties can be estimated. The air-dry densities of the 3 species studied are 536 kg/m^3 for box elder, 653 kg/m^3 for tree of heaven and 702 kg/m^3 for green ash.

The raw material base of pencil manufacturing is rather restricted. Basswood (*Tilia argentea*) is homogenous and smooth in texture, making its mechanical processing and surface treatment trouble-free. Our tests have shown that the impregnation capability of lower density linden is significantly better.

Only 18% of the low quality noble oak (*Quercus* spp.) boards can be used for defect-free lamella production. Taking into account the lower quality lamellae, their tests have shown that they can be used for the production of high-quality glued-laminated timber production with strength classes above D35.

Black locust (*Robinia pseudocacia*), sessile oak (*Quercus petraea*), beech (*Fagus sylvatica*), poplar (*Populus*), larch (*Larix decidua*) and spruce (*Picea abies*) were subjected to different test methods. Tensile strength, bending strength, compressive strength and hardness of the air-dry samples correlate well with the data from the literature. This confirms the correctness of our measurements. The examined properties consequently decreased with increasing moisture content between 12% moisture content and *FSP* for all six wood species, but the ratio of changes was various, depending both on the measurement method and on the species. The selected sample groups represented different groups of wood species, thus, these different sample groups will become better comparable.

This article was made in frame of the project TKP2021-NKTA-43 which has been implemented with the support provided by the Ministry of Culture and Innovation of Hungary from the National Research, Development and Innovation Fund, financed under the TKP2021-NKTA funding scheme. We express our special thanks to Imre Horváth for the preparation of the high-quality samples.

References

1. M. Báder, and R. Németh, Moisture-dependent mechanical properties of longitudinally compressed wood, *Eur. J. Wood Prod.* **77** (2019) 1009–1019. doi:10.1007/s00107-019-01448-1.
2. M. Bak, F. Molnár, R. Rákosa, Z. Németh, and R. Németh, Dimensional stabilization of wood by microporous silica aerogel using in-situ polymerization, *Wood Sci Technol.* **56** (2022) 1353–1375. doi:10.1007/s00226-022-01412-y.
3. F. Fodor, and M. Bak, Studying the Wettability and Bonding Properties of Acetylated Hornbeam Wood Using PVAc and PUR Adhesives, *Materials.* **16** (2023) 2046. doi:10.3390/ma16052046.
4. A. Ghavidel, A. Jorbandian, M. Bak, J. Gelbrich, J.J. Morrell, I. Sandu, and R. Hosseinpourpia, Degradation assessment of archaeological oak (*Quercus* spp.) buried under oxygen-limited condition, *Holzforschung.* **77** (2023) 198–207. doi:10.1515/hf-2022-0168.
5. F. Horváth, M. Bak, and M. Báder, Faanyagok gombaállóságának változása rostirányú tömörítést és további modifikációs eljárásokat követően, *Gradus.* **8** (2021) 59–65. doi:10.47833/2021.2.ENG.003.

6. M. Báder, and R. Németh, A review of wood compression along the grain—After the 100th anniversary of pleating, *Forests*. **14** (2023) 763. doi:10.3390/f14040763.
7. M. Bak, D. Takács, R. Rákosa, Z.I. Németh, and R. Németh, One-Step Process for the Fabrication of Hydrophobic and Dimensional Stable Wood Using Functionalized Silica Nanoparticles, *Forests*. **14** (2023) 651. doi:10.3390/f14030651.
8. Z.A. Köbölkuti, K. Cseke, A. Benke, M. Báder, A. Borovics, and R. Németh, Allelic variation in candidate genes associated with wood properties of cultivated poplars (Populus), *BioFut*. **70** (2019) 286–294. doi:10.1556/019.70.2019.32.
9. F. Fodor, M. Bak, A. Bidló, B. Bolodár-Varga, and R. Németh, Biological Durability of Acetylated Hornbeam Wood with Soil Contact in Hungary, *Forests*. **13** (2022) 1003. doi:10.3390/f13071003.
10. M. Báder, R. Németh, Á. Vörös, Z. Tóth, and A. Novotni, The effect of agroforestry farming on wood quality and timber industry and its supportation by Horizon 2020, *Agroforest Syst*. **97** (2023) 587–603. doi:10.1007/s10457-023-00812-8.
11. M. Bak, F. Molnár, and R. Németh, Improvement of dimensional stability of wood by silica nanoparticles, *Wood Material Science & Engineering*. **14** (2019) 48–58. doi:10.1080/17480272.2018.1528568.
12. F. Fodor, M. Bak, and R. Németh, Photostability of Oil-Coated and Stain-Coated Acetylated Hornbeam Wood against Natural Weather and Artificial Aging, *Coatings*. **12** (2022) 817. doi:10.3390/coatings12060817.
13. M. Báder, R. Németh, J. Sandak, and A. Sandak, FTIR analysis of chemical changes in wood induced by steaming and longitudinal compression, *Cellulose*. **27** (2020) 6811–6829. doi:10.1007/s10570-020-03131-8.
14. K. Cseke, Z.A. Köbölkuti, A. Benke, A. Rumi, M. Báder, A. Borovics, and R. Németh, Nemesnyár klónok faanyagtani jellemzőkhöz köthető génjeinek genetikai változatossága, *Erdészettudományi Közlemények*. **10** (2020) 5–16. doi:10.17164/EK.2020.001.
15. É. Lublóy, D.T. Mészáros, L.G. Takács, Z. Cimer, and H. Norbert, Examination of the fire performance of wood materials treated with different precautions, *J Therm Anal Calorim*. **148** (2023) 4129–4140. doi:10.1007/s10973-023-12050-2.
16. M. Bak, and R. Németh, Effect of different nanoparticle treatments on the decay resistance of wood, *BioResources*. **13** (2018) 7886–7899.
17. M. Báder, R. Németh, and J. Konnerth, Micromechanical properties of longitudinally compressed wood, *Eur J Wood Prod*. **77** (2019) 11. doi:10.1007/s00107-019-01392-0.
18. FAO, State of the World's Forests 2009, Rome, Italy, 2009.
19. F. Steierer, Current wood resources availability and demands national and regional wood resource balances EU/EFTA countries, United Nations, Geneva, Switzerland, 2010.
20. UNECE-FAO, FAO: The European Forest Sector Outlook Study II 2010-2030, United Nations, United Nations Economic Commission for Europe, Food and Agriculture Organization of the United Nations, Geneva, 2011.
21. P.R.J. Shukla, R. Skea, K. Diemen, Ø. Calvin, F. Christophersen, and J. Creutzig, eds., Meeting report of the intergovernmental panel on climate change expert meeting on mitigation sustainability and climate stabilization scenarios, in: IPCC Working Group III Technical Support Unit, London, UK, Imperial College London, 2017.

22. M. Hanewinkel, D. Cullmann, M.J. Schelhaas, G.J. Nabuurs, and N.E. Zimmermann, Climate change may cause severe loss in economic value of European forestland, *Nat. Clim. Chang.* **3** (2013) 204-207,. doi:10.1038/nclimate1687.
23. V. Altieri, S. Franco, F. Lombardi, P.A. Marziliano, G. Menguzzato, and P. Porto, The role of silvicultural systems and forest types in preventing soil erosion processes in mountain forests: A methodological approach using cesium-137 measurements, *J. Soils Sediments.* **18** (2018) 1-10,. doi:10.1007/s11368-018-1957-8.
24. K. Updegraff, M.J. Baughman, and S.J. Taff, Environmental benefits of cropland conversion to Hybrid poplar: economic and policy considerations, *Biomass Bioenergy.* **27** (2004) 411–428. doi:DOI:10.1016/j.biombioe.2004.05.002.
25. M. Kiaei, Radial variation in wood static bending of naturally and plantation grown alder stems, *Cellulose Chem. Technol.* **47** (2013) 339–344.
26. G. Ónodi, Az idegenhonos, illetve inváziós fafajok élőhelyformáló hatásai, *Erdészettudományi Közlemények.* **6** (2016) 101–113. doi:10.17164/EK.2016.008.
27. C. Harvald, and P.O. Olesen, The variation of the basic density within the juvenile wood of Sitka spruce (Picea sitchensis,) Scand, *J. For. Res.* **2** (1987) 525–53. doi:10.1080/02827588709382488.
28. MSZ 45:2022 Fűrészrönk. Követelmények és megjelölés (Sawlogs. Requirements and marking)., 2022.
29. EN 408:2010+A1:2012: Timber structures - Structural timber and glued laminated timber - Determination of some physical and mechanical properties, 2012.
30. ISO 13061-06: Physical and mechanical properties of wood - Test methods for small clear wood specimens - Part 06: Determination of ultimate tensile stress parallel to grain, 2014.
31. ISO 13061-03: Physical and mechanical properties of wood - Test methods for small clear wood specimens - Part 03: Determination of ultimate strength in static bending, 2014.
32. ISO 13061-17: Physical and mechanical properties of wood - Test methods for small clear wood specimens - Part 17: Determination of ultimate stress in compression parallel to grain, 2017.
33. MSZ 6786-11: Test methods for wood materials. Determination of hardness, 1982.
34. EN 338:2016 Structural timber. Strength classes, 2016.
35. S. Molnár, Knowledge of wood (Faanyagismeret), Mezőgazdasági Szaktudás Kiadó, Budapest, Hungary, 2004.

The Effect of Wooden Building Materials on the Indoor Air Quality of Houses

Csilla Patkó[1], Péter Adamik[1], and Zoltán Pásztory[1*]

[1] University of Sopron, 9400 Sopron, Hungary

Abstract. This work focuses on researching wood as a low-emitting building material. Many studies showed that by regulating the emission of adverse chemicals from pressed wood products, they could be recycled as low-emitting materials. The presented case study, shows the measurements of the indoor air quality in a passive, low-energy, wooden, light frame house. Concentrations of Volatile Organic Compounds (VOC) and formaldehyde were measured and compared with data from other case studies made in this subject and with the current Hungarian and international regulations of adverse chemicals. The main VOC materials which concentrations were significant in all measurements were terpenes (limonene, alpha-pinene, 3-carene), aliphatic-hydrocarbons and aldehydes (acetic acid, 2-Methoxy-1-ethylmethyl acetate). The comparison showed that the concentrations of benzene, toluene, naphthalene, formaldehyde, and styrene were not deviant from the average values measured by other case studies. The observations showed that the main influencing factors of indoor air quality were the changes in indoor temperature, relative humidity, air exchange rate, and human activities. The conclusion is that the wooden materials applied in the house are low-emitting materials and do not pose a health risk for people.

1 Introduction

The global climate change and increasing energy requirements have led to the development of airtight, energy-saving buildings with very low air exchange rates. Because of the cheap mass-production the most common building materials are synthetic materials. The houses may become more and more comfortable and automatized, still the question arises: Do these technical improvements provide a healthy living-space for people?

We spend 80-90% of our lives inside buildings [1]. Environmental health researchers have been investigating intensively the indoor air quality of newly built homes, existing homes and converted homes [2] [3] [4]. The guidelines of WHO (World Health Organisation) categorize the harmful substances emitted by the sources [5]. In Germany the Committee for the Health Assessment of Building Materials first defined VOCs from the emission of building products in 2000. Now more than 180 different VOCs can be defined [6]. There are some countries where the regulations of the indoor air quality define the building materials and the concentrations of adverse chemicals emitted from them. In the European Directive

* Corresponding author: pasztory.zoltan@uni-sopron.hu

emphasized, that the buildings cannot pose a health risk to the residents through the emitted adverse chemicals [7].

With the right choice of building materials and technologies the sources of adverse chemicals can be avoided. Wood is one of the most environment-friendly raw materials [8] [11]. Wood cells contain high amount of carbon absorbed from the CO2 component of the air. Engineered wood products contain chemical materials in form of added adhesives or as surface treatment etc. These chemicals have significant emissions of volatile organic compounds (VOCs) that could cause health risk especially in indoor spaces [9] [10].

There are approaches made to develop ecological technologies in manufacturing and treatment of wooden materials [12-16]. There are several studies about the emission of VOCs and formaldehyde from wood-based composites, such as MDF, plywood, particleboards, etc. Ze-Li showed in their study that the most emitted VOCs were terpenes, aldehydes, and aromatics. Pine wood plank had the highest VOCs (approximately 900 μg/m3). Fresh particleboard showed the second highest value with 450 μg/m3 [17]. Another study compared the emissions of VOCs from air-dried pine wood and heat-treated pine wood. Significant chemical changes have occurred, and volatile monoterpenes and other low molecular weight compounds have evaporated from the wood during the heat-treatment process when compared to air dried wood [18]. The emission of wood-based materials depends on the type of wood. Hard wood emits high concentrations of acetic acid and formic acid, while soft wood emits more terpenes (alpha-pinene, 3-carene, etc.) [19].

Song-Yung measured the emissions of particleboard from recycled wood-waste chips using polymeric 4,4-methylenediphenyl isocyanate (PMDI) and phenol-formaldehyde (PF) resins. PMDI is one of the most used aromatic isocyanates in the world. 5 million of tones of PMDI are produced yearly. It doesn't pose a health risk, but it can cause irritation [20].

In other studies, there were used natural resins instead of UF or PF at the manufacturing of engineered wood. Sumin Kim used natural tannin as adhesive in wood-based flooring. They had to add PVAc to increase the bonding strength of tannin. The formaldehyde emission was lowered even by adding PVAc (Polyvinyl acetate), and more greatly reduced when UV curable urethane acrylate was coated [21]. The volcanic pozzoloan is another material added as scavenger to engineered wood, which reduced the emissions of formaldehyde [22]. In a newly-built wooden house the concentrations of the emitted adverse chemicals can be reduced with the "bake out" process. The inside temperature is raised to 32-40°C and kept it for a week. The extent of the reduction of adverse chemicals in the indoor air is about 60-90% [23].

There are lots of examinations made in test chambers measuring the emission of chemicals from the single building products. But still there are less study showing measurements of adverse chemicals in the indoor air of wood buildings [24].

The presented case study shows the measurements of the indoor air quality in a passive, low-energy, wooden, light frame house. Concentrations of Volatile Organic Compounds (VOC) and formaldehyde were measured and compared with data from other case studies made in this subject and with the current international regulations of adverse chemicals.

2 Measurement

The sampling was performed in a newly built, low-energy, frame-house. Description sees in our former paper [25]. Six measurements were taken in different phase of building process throughout the period of a year. The results of the measurements showed the changing concentrations of adverse chemicals (Table 1).

Table 1. Comparison of the measured values with the mean values related to Middle-European case-studies.

Measure-ments	1.	2.	3.	4.	5.	6.	Mean value Middle Europe
	($\mu g/m^3$)						
Benzene	3.11	2.79	0	1.66	2.35	1.65	3.1
Toluol	11.6	5.09	6.13	19.07	38.4	4.81	20.6
Styrol	3.06	1.32	0	4.98	147	7.93	1.1
Formaldehyde	34	11	130	49	99	32	29.8
Naphtalene	0	0	0	1.81	1.15	0	1.1
Limonene	64.7	21	15.8	59.7	289	30.2	17.2
Alpha-pinene	297	170	300	300	2710	284	12.9

By summing the measured concentrations of VOCs, we get the TVOC values. These values are plotted in Fig.1. The graph clearly shows that TVOC concentrations were influenced by indoor temperature, indoor relative humidity, and air exchange rate. Human activities and the furniture had the greatest impacts on the concentrations. The TVOC value of the first measurements was 2150 μg/m3, the last was 1190 μg/m3. After the first measurement the air-heating was turned on, and the indoor temperature was not increased, this is the reason of the decreasing values. Before the 4. measurement the floor-heating was turned on, the indoor temperature lifted to 21°C. This must have caused the increased concentrations. This assumption of ours seems to be confirmed by study of Young [28]. During June the indoor temperature raised above 24°C which indicated the significant change in the concentrations. In this period were the rooms furnished, which could have affected the increasing values as well. Probably the frequent natural ventilation indicated the lowering of high VOCs concentrations.

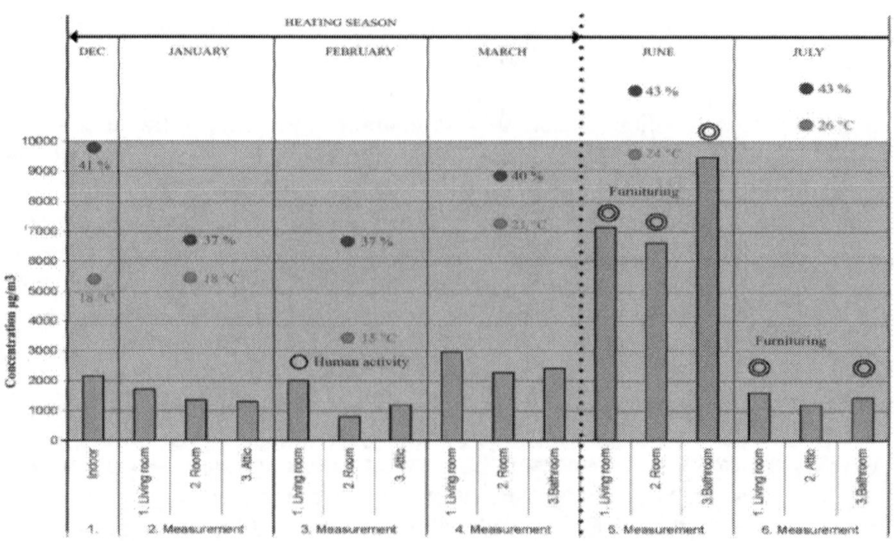

Fig. 1. TVOC values of the measurements. The red dots show the indoor temperature, the blue dots show the indoor relative humidity.

The concentrations of benzene (3.5 µg/m3) were not deviant from the average values measured in other case studies, and from the current Hungarian and international regulations of adverse chemicals (Fig. 2). They did not pose a health risk. The concentration of toluol was between 2.9 – 38.4 µg/m3. The increase was affected by indoor temperature, indoor relative humidity, and the presence of furniture in the room. The highest concentration exceeded the maximal value given by WHO (20 µg/m3).

The concentration of formaldehyde with the value of 130 µg/m3 did overpass the values of the current international regulations by the WHO (100 µg/m3) during the third measurement, but it was just temporary (Fig. 3). The main sources were the wooden materials cut into pieces in the living room. By the following measurements the concentration of formaldehyde was slightly increased because of the rise of the indoor temperature.

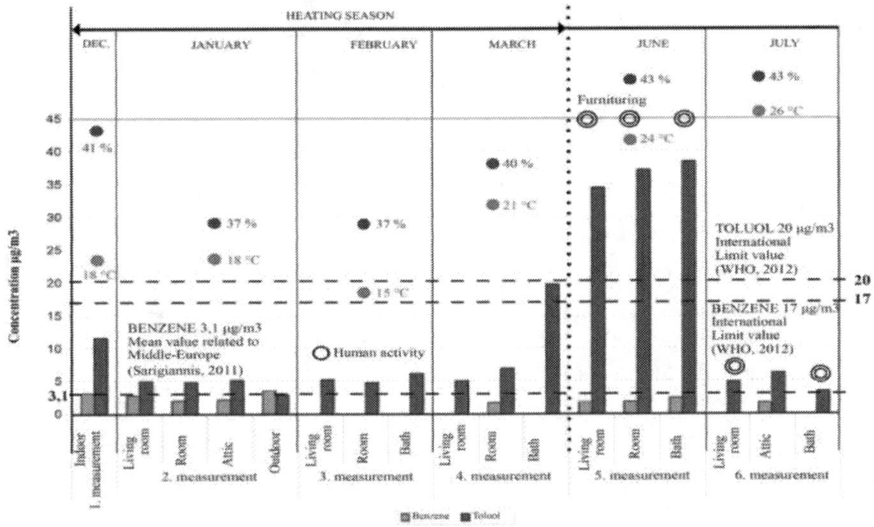

Fig. 2. Benzene and toluol values of the measurements. The red dots show the indoor temperatures, the blue dots show the indoor relative humidity.

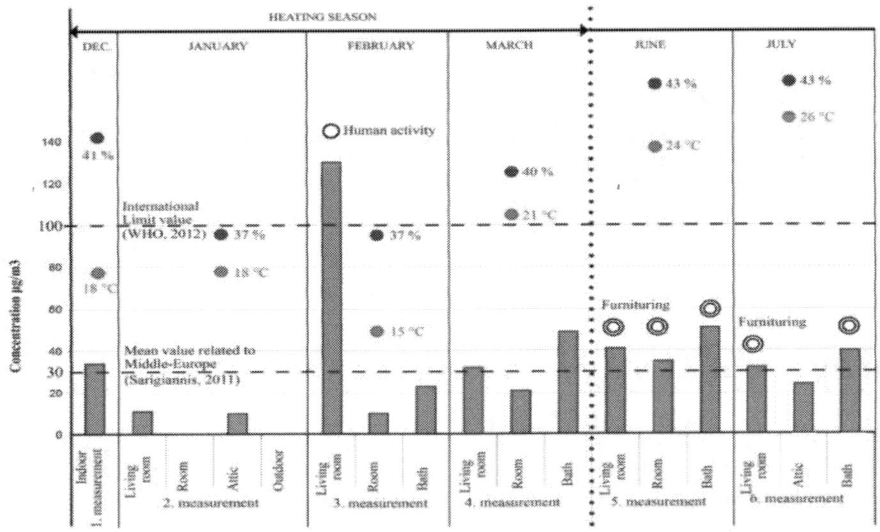

Fig. 3. Formaldehyde values of the measurements. The red dots show the indoor temperatures, the blue dots show the indoor relative humidity.

The rate of the wooden clothing in each room had a great influence on the concentrations of alpha-pinene and 3-carene. These chemicals are not toxic, therefore they did not pose a health risk. Fig.4. shows the influence of indoor temperature, indoor relative humidity, and the presence of furniture in the rooms on the concentrations.

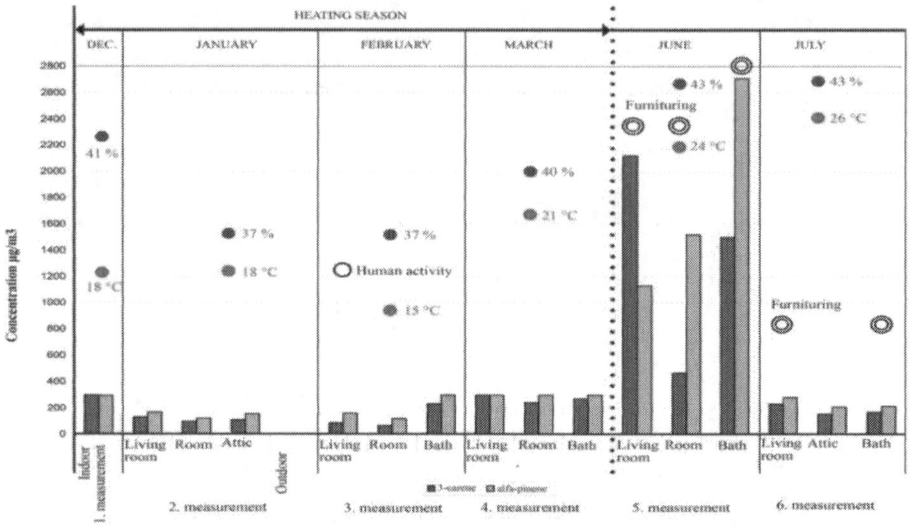

Fig. 4. 3-carene and alpha-pinene values of the measurements. The red dots show the indoor temperatures, the blue dots show the indoor relative humidity.

3 Comparison with other studies

During literature survey we have found around 200 papers regarding in-situ measurements of IAQ in buildings, reviews and chamber test methods of building materials. After sampling the measurement data we compared some of them with our values. But it still shows a very chaotic picture about the guidelines of measurements, which makes it hard to evaluate them. Each country has different regulations of maximal exposures of adverse chemicals and different measuring methods as well.

For a detailed analysis a further research work is needed.

The case studies were categorized into three main groups:

- Conventional, newly-built or renovated houses (before moving in and after moving in)
- Conventionally-built houses
- Houses with low-emitting building materials (before moving in and after moving in)

The strategy of the comparison was based on the maximal and minimal TVOC values measured in our study. In Table 2 are the chemicals chosen to be compared.

3.1 Conventionally built and renovated houses

Rothweiler measured in their study the IAQ of newly-built or renovated apartments. Their results showed an elevated percentage of aldehydes, ketons, and alcohols. They suspected that the high concentrations and the lack of ventilation can cause - immediately after moving - in symptoms of head-ache, irritation of the eyes. The presence of furniture and human activity cause an additional rise of the concentrations table 2 [26].

Seung H. examined the indoor air quality of 107 newly-built apartments in Korea at the pre-occupancy stage. The dominant VOCs were: formaldehyde, aromatic hydrocarbons, alcohols, terpenes, and ketons. The high concentration of formaldehyde could be caused by not following the Korean Ministry of Environment guidelines for formaldehyde emissions. The main sources were suspected to be wood panels, vinyl floor coverings, adhesives and paints table 2 [27].

The conclusion is that our measured values are mostly the lowest, with the exception of limonene. That is the effect of the high percentage of the wooden materials in the house.

3.2 Conventionally built houses

Kostiainen quantified 48 compounds on the basis of their prevalence, toxicity, carcinogenicity and mutagenicity in 50 normal houses. They have measured high concentrations of VOCs in normal houses, which suspected, that not only VOCs are affecting the indoor air quality [29].

Tuomainen measured the indoor air quality of normal houses, and found that the concentration of TVOC were significantly higher before the residents moved in because of the deficient ventilation and the high-emitting building materials [30].

Compared to these two case studies the minimal values from our measurements were lower, but the maximal values were significantly higher (Table 2).

Table 2. Comparison of values of VOCs and TVOC measured in conventionally built houses with values from relevant case studies found in the literature.

Components	Measure values Patkó et al. (2013)		Case studies from literature			
			KOSTIAINEN (1995)		Tuomainen et al. (2000)	
	Min. value	Max. value	Min. value	Max. value	Before moving in	After 5 months
Benzene	1.65	3.11	0.21	38.55	-	-
Toluol	4.81	38.4	0.6	70.36	-	84
Styrol	1.32	147	0.08	3.87	-	-
Formaldehyde	11	130	-	-	12.0-27.0	5.0-19.0
Naphtalene	1.15	1.81	0	1.63	-	-
Limonene	15.8	289	1.34	51.56	-	19
Alpha-pinene	170	300	1.05	36.41	-	51
TVOC	**795**	**7130**	**40.84**	**235.85**	**1290.0-9580.0**	**170.0-1335.0**

3.3 Houses with low emitting building materials

Tuomainen investigated the IAQ of houses built by following the instructions of the Finnish Classification of Indoor Climate, Construction and Finishing Materials. The ventilation system was kept at a high capacity for one week after its completion before occupants moved in. Indoor air parameters were measured in one apartment on each floor before occupants moved in (2 measurements) and after a 5-month occupancy [30].

In the study of Guo a low volatile organic compound (VOCs) emission house was investigated. The construction materials used in the house were tested in an environmental chamber and low VOC emission materials were then selected. The ventilation was maximized by the right placement of the windows. The TVOCs concentrations measured in the house ranged from non-detectable to 43 (μg/m3). This study confirmed that for controlling IAQ was pollution prevention very effective and the next most important was the design of ventilation rates to handle uncontrollable sources [31].

The volatile organic compounds (VOCs), formaldehyde, and ammonia concentrations as well as temperature, relative humidity, and the air exchange rate were determined right after the finish and after 6 and 12 months. 240 VOCs were detected, from which xylol and alpha-pinene had the highest concentrations. After six months the values of TVOC were significantly decreasing. The concentrations of ammonia and formaldehyde showed seasonal variations, i.e., higher concentrations were measured in summer [32].

Compared to these two case studies the minimal values from our measurements were lower, but the maximal values were significantly higher (Table 2 and Table 3).

Table 3. Comparison of values of VOCs and TVOC measured in ecologically built houses with values from relevant case studies found in the literature.

	Rothweiler et al. (1992)	Seung H. et al. (2012)	Case studies from literature						
			Tuomainen et al. (2000)			Guo et al. (2003)	Jarnstrom et al. (2006)		
Components	Avg. value	Avg. value	Before aeration	After 1 week	After moving in	After 1 year	After finish	After 6 months	After 1 year
Benzene	20	3.9	-	-	-	-	0	0	3
Toluol	9741	184	-	19	12	-	-	-	-
Styrol	167	2.7	-	-	-	-	3	2	3
Formaldehyde	1200	62	1.0 - 20.0	6.00 - 15.00	2.00 - 21.0	-	19	21	26
Naphtalene	219	0.8	-	-	-	-	-	-	-
Limonene	224	4.3	-	140	23	-	12	10	12
Alpha-pinene	867	9	89	46	-	-	61	37	35
TVOC	**31696**	-	**210-1800**	**100 - 1100**	**61 - 410**	**0 - 43**	**780**	**329**	**247**

4 Conclusion

The comparison showed that the concentrations of benzene, toluene, naphthalene, formaldehyde and styrene were not deviant from the average values measured by other case studies. The observations showed that the main influencing factors of indoor air quality were the changes in indoor temperature, relative humidity, air exchange rate, and human activities. The conclusion is that the wooden materials applied in the house are low-emitting materials and do not pose a health risk for people.

The measurements of indoor air quality showed a great number of new information about the correlation of building materials, the use of the building, and indoor air quality. They are providing the designers and the building occupants with essential information about the state of the building and the health impacts. They can be used in many types of the architectural processes, e.g. in building diagnostics for renovations, in measuring the health impacts of a newly built house, and in finding the main sources of adverse chemicals of buildings with "sick building syndrome".

By choosing the right building materials (low-emitting materials) and technologies can these adverse impacts on the indoor air quality be avoided. For this, there must be a regulated and up-to-date database of low-emitting materials accessible for designers and building contractors, to create with breathing structures an environmentally conscious, healthy living-space. More in-situ measurements are needed, to get a wider view of the health state of buildings and how they affect our health.

References

1. WHO: *Guidelines for indoor air quality: selected pollutants*, Denmark: WHO Regional Office for Europe, ISBN 978 92 890 0213 4 (2010)
2. T. Salthammer, S. Mentese, R.Marutzky, *Formaldehyde in the indoor environment*. Chemical Reviews **110**, (pp. 2536 – 2572), (2010)
3. P. Wolkoff, *How to measure and evaluate volatile organic compound emissions from building products*. A perspective. Science of the Total Environment, **227**, (pp. 197–213.) (1999)
4. B. Holmberg, P. Lundberg, *Benzene: standards, occurrence, and exposure*. Am J Ind Med; **7**:375–83. (1985)
5. WHO. *Air quality guidelines, global update 2005, particulate matter, ozone, nitrogen dioxide and sulphur dioxide*, Denmark: WHO Regional Office for Europe, 2006, ISBN 92 890 2192 6 (2006)
6. AgBB, Updated List of LCI values 2012 in Part 3, Health-related Evaluation Procedure for Volatile Organic Compounds Emissions (VOC and SVOC) from Building Products, (2012)
7. A. Katsoyiannis, P. Leva, J. Barrero-Moreno, D. Kotzias, *Building materials. VOC emissions, diffusion behaviour and implications from their use*. Environmental Pollution **169** (pp. 230 – 234) (2012)
8. S. Molnár *Forest and Wood Utilization in Hungary, Erdő - fa hasznosítás Magyarországon*, NyME Erdészeti Tudományos Intézet, (p. 344) (2005)
9. H. Guo, F. Murraya, S.-C. Lee, *Emissions of total volatile organic compounds from pressed wood products in an environmental chamber*. Building and Environment **37**, (pp. 1117 – 1126) (2002)
10. EPA (Environmental Protection Agency) *Sources and factors affecting indoor emissions from engineered wood products: summary and evaluation of current literature.* EPA-600=R-96-067. North Carolina: Research Triangle Park, (1996)
11. A. Petersen, B. Solberg, *Environmental and economic impacts of substitution between wood products and alternative materials: a review of micro-level analyses from Norway and Sweden*. Forest Policy and Economics **7**, (pp. 249– 259) (2005)
12. P. Takáts *Formaldehyde content of wood-based composite boards; Faalapú kompozit lemezek formaldehid tartalma I.* Magyar Asztalos IV: **5** (pp. 48–49) (1994)
13. P. Takáts *Formaldehyde content of wood-based composite boards; Faalapú kompozit lemezek formaldehid tartalma. II.* Magyar Asztalos IV: **7**, (p. 47) (1994)
14. Takáts P. Emission threshold limits of composite panels; Kompozit lemezek emissziós határértékei. Magyar Asztalos IV: **9**, (p. 56) (1994)
15. P. Takats, A. Winkler, L. Albert, Zs. I. Nemeth *Formaldehyde Contents of Particleboard: Comparing WKI–Method and with FESYP Perforator Test Method*,

„5th International, Jubilee Conference on Role of Formaldehyde in biological systems, October 9−13. Sopron, Hungary, (pp. 54-58) (2000)

16. T. Alpár, R. Németh, P. Takats, *Renewable composites, materials, R+D+I on the Wood Science Faculty; Megújuló Kompozitok, anyagok, K+F+I a Faipari Mérnöki Karon,* Inno-Lignum.09.06-08. (2012)
17. Ze-Li, Q., Fei-Bin, W., Jian-Zhang, L., Takeshi, F. Assessment on emission of volatile organic compounds and formaldehyde from building materials. Composites: Part B **49**, 36–42 (2013)
18. A. Manninena, P. Pasanenb, J.K. Holopainen, *Comparing the VOC emissions between air-dried and heat-treated Scots pine wood.* Atmospheric Environment **36**, (pp. 1763–1768) (2002)
19. M. Risholm-Sundman, M. Lundgren, E. Vestin, P. Herder *Emission of acetic acid and other volatile organic compounds from different species of solid wood*, Holz als Roh und Werkst. **56**, (pp. 125–129) (1998)
20. W. Song-Yung, Y. Te-Hsin, L. Li-Ting, L. Cheng-Jung, T. Ming-Jer *Properties of low-formaldehyde-emission particleboard made from recycled wood-waste chips sprayed with PMDI/PF resin*, Building and Environment **42**, (pp. 2472–2479) (2007)
21. K. Sumin, *Environment-friendly adhesives for surface bonding of wood-based flooring using natural tannin to reduce formaldehyde and TVOC emission*, Bioresource Technology **100**, (pp. 744–748) (2009)
22. K. Sumin, *The reduction of indoor air pollutant from wood-based composite by adding pozzolan for building materials*, Construction and Building Materials **23**, (pp. 2319–2323) (2009)
23. K. Sumin, C. Yoon-Ki, P. Kyung-Won, K. Jeong Tai *Test methods and reduction of organic pollutant compound emissions from wood-based building and furniture materials*, Bioresource Technology **101**, (pp. 6562–6568) (2010)
24. C.J. Weschler, *Changes in indoor pollutants since the 1950s*, Atmospheric Environment **43**, (pp. 153–169) (2009)
25. Cs. Patkó, *Evaluation of volatile organic components of indoor air of a newly-built wooden frame house in the last four seasons*. Conference Proceedings Part (A) 4th INTERNATIONAL CONFERENCE “TO PROTECT OUR GLOBAL ENVIRONMENT FOR FUTURE GENERATIONS” ICEEE-, 20 – 21 November 2013, Óbuda University Budapest, Hungary, (pp. 229-237) ISBN: 978-615-5018-93-0 (2013)
26. H. Rothweiler, P. Wager, C. Schlatter, *VOLATILE ORGANIC COMPOUNDS AND SOME VERY VOLATILE ORGANIC COMPOUNDS IN NEW AND RECENTLY RENOVATED BUILDINGS IN SWITZERLAND*. Atmospheric Environment Vol 26A No **12**, (pp 2219-2225) (1992)
27. H. Seung, K. Wan, *Volatile organic compound concentrations, emission rates, and source apportionment in newly-built apartments at pre-occupancy stage*. Chemosphere **89** (pp. 569–578) (2012)
28. L. Young-Kyu, K. Hyun-Joong, *The effect of temperature on VOCs and carbonyl compounds emission from wooden flooring by thermal extractor test method*. Building and Environment **53**, (pp. 95 – 99) (2012)
29. R. Kostiainen, *Volatile organic compounds in the indoor air of normal and sick houses*. Atmospheric Environment, Vol. **29**, No. 6, (pp. 693-702) (1995)

30. M. Tuomainen, A-L. Pasanen, A. Tuomainen, J. Liesivouri, P. Juvonen, *Usefulness of the Finnish classification of indoor climate, construction and finishing materials: comparison of indoor climate between two new blocks of flats in Finland* Atmospheric Environment **35**, (pp. 305-313) (2001)
31. H. Guo, F. Murray, S.C. Lee, *The development of low volatile organic compound emission house—a case study*. Building and Environment **38**, (pp. 1413 – 1422) (2003)
32. H. Järnström, K.Saarela, P. Kalliokoski, A.-L. Pasanen, *Reference values for indoor air pollution concentrations in new residential buildings in Finland*. Atmos. Environ. **40**, (pp. 7178–7191) (2006)

Reducing Indoor Air Pollution through Personalized Ventilation for Occupants in Office Environments and Confined Spaces

Florin Bode[1,2]* , *Ilinca* Nastase[2], *Paul* Danca[2,3]

[1]Mechanical Engineering Department, Faculty of Automotive, Mechatronics and Mechanics, Technical University of Cluj-Napoca, Cluj-Napoca, Romania.
[2]CAMBI Research Center, Faculty of Building Services, Technical University of Civil Engineering Bucharest, Bucharest, Romania
[3]National Institute for R&D in Electric, Engineering ICPE-CA, Bucharest, Romania.

Abstract. With the increasing focus on indoor environmental quality, driven by the growing amount of time people spend in enclosed spaces, this study presents an approach to enhancing air distribution in office environments and confined spaces. A novel low-induction air diffuser is designed to deliver fresh, clean air in close proximity to occupants while maintaining their thermal comfort. Clean, unpolluted air is pivotal to healthy and productive workplaces. Yet, this paper underscores the importance of not sacrificing thermal comfort in the pursuit of improved indoor air quality. Inadequate thermal comfort may lead occupants to deactivate ventilation systems, negating the benefits of improved air quality. Inefficient temperature control can also result in discomfort, distractions, and reduced productivity. The innovative low-induction air diffuser resolves this issue, enhancing air quality near occupants without causing thermal discomfort. By directing air gently and efficiently, this solution is prepared to transform personalized ventilation systems, mitigating the discomfort associated with traditional jet flows while delivering high-quality breathable air. This research serves as a bridge between improved indoor air quality and thermal comfort, for office environments. It introduces a practical, energy-efficient solution that satisfies the core requirements of a healthy workspace—clean air and comfortable conditions.

1 Introduction

Indoor air quality (IAQ) in buildings is a major concern that demands immediate attention when evaluating the well-being and productivity of individuals in the workplace. The fundamental concept underpinning mixing ventilation, which is the most popular ventilation method in buildings, is the blending of supplied air with the pre-existing air within an enclosed space. This approach involves delivering clean air from diffusers situated at a distance from occupants. IAQ it can be profoundly influenced by passive contaminants, such

* Corresponding author: florin.bode@termo.utcluj.ro

as gases and airborne particulate matter including both indoor-generated particles (PM2.5, PM10) and outdoor-generated traffic-related particles [1-3]. These pollutants present a substantial risk to human health as they can affect occupants and lead to respiratory problems like asthma [4] or respiratory allergies [5], or can lead to cardiovascular diseases [6], or eyes, nose and throat irritations [7], thereby contributing to serios health problems for the occupants. Moreover, in office spaces, as opposed to other indoor environments, the vulnerability of occupants to contamination is exceedingly high, both through direct inhalation and indirectly via the deposition of particles on surfaces, followed by hand-to-mouth contact [5]. Therefore, ensuring high-quality breathable air at the same time without affecting the thermal comfort of the occupants within office spaces is of utmost importance, making the choice of a suitable air distribution system essential.

Personalized ventilation (PV) has been introduced as a solution to improve IAQ capable of satisfying both of these specific requirements [8].

Regarding air quality, personalized ventilation systems can contribute significantly to improving breathable air quality in office spaces. These systems can be equipped with advanced air filtration technologies and sensors to detect and respond to changes in air quality. By ensuring that each occupant receives a fresh and purified air supply, the risk of exposure to pollutants and allergens is greatly reduced. This, in turn, mitigates the potential health issues associated with poor indoor air quality, including respiratory problems and allergies.

In the context of thermal comfort, traditional HVAC systems often provide uniform temperature control for entire spaces, which might not align with the individual preferences of occupants. Personalized ventilation, on the other hand, allows for customized control, tailoring the thermal environment to each person's needs. By providing adjustable airflow and temperature settings at the individual level, it offers the promise of a more comfortable working environment, thus enhancing the overall well-being and productivity of office occupants.

PV presents an advanced capability to supply occupants in office environments and confined spaces with clean, fresh air precisely within their breathing zone [9-12]. Simultaneously, it tailors the microclimate to harmonize with individual thermal preferences. This approach enhances the overall quality of the indoor environment by delivering personalized comfort and air quality to occupants. Extensive research has been conducted to explore various configurations of PV systems, including ceiling and desk air terminal diffusers, with the objective of achieving thermal comfort and optimal air quality at varying levels of efficacy.

For example, in [12], the researchers conducted experiments in which they assessed the air quality of inhaled air by utilizing tracer gas. Their study investigated various types of air terminal devices and yielded intriguing results. Specifically, it was observed that a vertical desk grill demonstrated the highest air quality throughout the room space, whereas a round movable panel achieved the highest air quality within the occupant's breathing zone [13, 14].

In a related study, as detailed in [15], researchers conducted a comprehensive field study that focused on assessing IAQ in office environments where both displacement ventilation and personalized ventilation systems with optional window operation were employed. Their investigation uncovered noteworthy insights into the IAQ dynamics. Specifically, it was observed that under warmer conditions, opting for a higher PV flow rate operation led to the attainment of acceptable IAQ levels, highlighting the adaptability and efficacy of PV systems in creating comfortable working environments.

Furthermore, in a separate field study in an office space, as referenced in [11] the authors evaluated the impact of five different PV air terminal designs on the perceived air quality at individual workstations. Their findings revealed a notable improvement in perceived air quality across all workstations with the implementation of these PV systems. This underscores the capacity of PV solutions to enhance occupant comfort and underscores their potential as a viable strategy for optimizing IAQ in office spaces.

Personalized ventilation systems aim to position supply air in close proximity to the occupant. In contrast to traditional ventilation systems, where as little as 5% of supplied air reaches the occupant, PV systems are capable of delivering a substantial 50% or more of the supply air directly to individuals in the space [16]. The objective of this paper is to develop a novel air diffuser with low induction using numerical simulation, capable of delivering fresh and clean air to the proximity of the occupant face, improving the IAQ nearby occupant, in a way to not affect his thermal comfort.

2 Materials and methods

2.1 Analysed geometry

The chosen solution resulted from the work carried out in the frame of scientific research grants [17-19] and aims to provide fresh and clean air as close as possible to the occupants, yet without affecting in a negative way the thermal comfort of the users. The solution consists of a steerable diffuser which will deliver fresh air in the proximity to the occupant's face.

Characteristic of this type of spherical diffuser with low induction lobed orifice for personalized ventilation **Fig. 1**a is the fact that the jet of air instilled through the diffuser will entrain a much smaller amount of air than in the case of a classic circular jet due to the specific shape of the orifices geometry [10, 20, 21].

In our setup, a low-induction spherical diffuser was incorporated into a virtual environment **Fig. 1**b. Within this virtual environment, a seated virtual manikin was positioned in front of a desk with a notebook [22, 23]. The shape of the humanoid manikin is similar with the real human with a few simplifications which will not affect the results [24, 25]. Two air diffusers were employed and positioned at the top of the notebook screen as depicted in **Fig. 1**b.

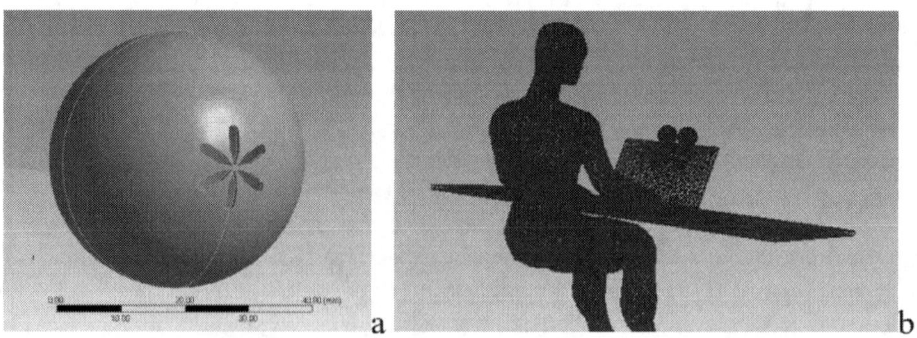

Fig. 1. a. Considered air diffuser geometry b. Entire geometry setup

2.2 Numerical grid

The numerical grids setup was performed based on our previous experience. A study of solution independence was conducted depending on the number of grid elements used (0.16 million, 0.85 million, 2 million, 5 million, 8 million). The numerical grid chosen for the numerical study was that with 5 million elements (**Fig. 2**) because of the small differences in numerical results between the 5 and 8 million element mesh.

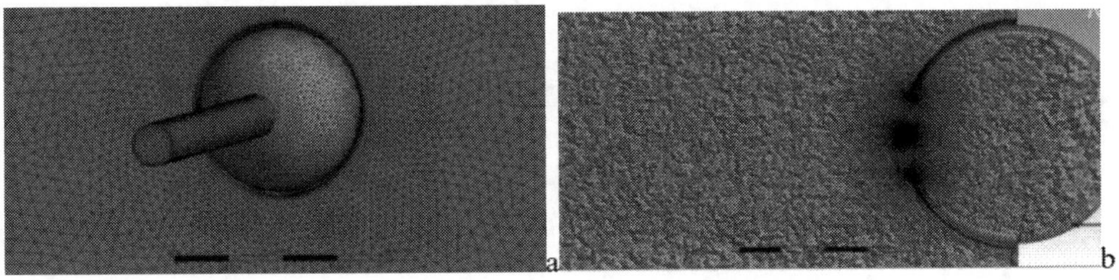

Fig. 2. Details with the 5 million element numerical grid

The mesh consisted of tetrahedral elements which play a crucial role in Computational Fluid Dynamics (CFD). These elements, characterized by their pyramidal shape, offer several advantages in mesh generation for CFD analyses. First and foremost, tetrahedral elements are highly versatile and can be effectively employed in complex geometries and their ability to adapt to irregular shapes and refine mesh in regions of interest contributes to enhanced accuracy and resolution in capturing flow phenomena. Tetrahedral elements are particularly valuable when dealing with unstructured meshing, allowing for greater flexibility in defining mesh densities where they are needed most [26]. This adaptability is especially beneficial when simulating complex and dynamic fluid behaviours making them essential for modelling the airflow through the lobed shape of the air diffuser or the natural convection in the proximity of the human shaped virtual manikin.

2.3 Numerical configuration

The flow rate was imposed considering the recommendations from [16]. The total airflow rate was 10 l/s (36m^3/h). This was the main reason why the proposed solution consists of two novel air diffusers with low induction.

The numerical simulation CFD study was conducted on this proposed geometry. Previous experience related to the study of jets was used to perform the numerical simulations. The turbulence model used was SST k-ω.

Utilizing the Shear-Stress Transport (SST) k-ω turbulence model for simulations in the context of moderate Reynolds numbers within the personalized ventilation zone offers several distinct advantages. The SST k-ω model combines the capabilities of the k-ε and k-ω models, making it well-suited for capturing a wide range of turbulent flow behaviors with precision. Here, we'll delve into the specific benefits of employing this model for personalized ventilation simulations.

One of the primary advantages of the SST k-ω model is its reliability in predicting turbulence in moderate Reynolds number flows. In personalized ventilation, where the airflow is often characterized by lower velocities and shear, this model excels at capturing the intricacies of turbulence, ensuring that the simulation results are both accurate and trustworthy.

Moreover, the SST k-ω model is known for its robustness and ability to handle adverse pressure gradients, a common occurrence in personalized ventilation scenarios where the airflow can be quite complex [26]. This robustness ensures that the simulations remain stable and converge effectively, even in regions with rapidly changing flow characteristics like in the case of our setup.

The model's effectiveness in resolving near-wall flows is particularly advantageous in personalized ventilation simulations. In the vicinity of walls or surfaces where occupants are situated, the flow patterns are critical for understanding exposure to contaminants or improving thermal comfort. The SST k-ω model excels in these scenarios, offering insights into the near-wall turbulence behavior that other models may struggle to provide.

 https://doi.org/10.1051/e3sconf/202451404004

Additionally, the SST k-ω model's compatibility with unstructured grids facilitates meshing in intricate geometries, which is often the case in personalized ventilation scenarios where the airflow paths can be irregular and highly localized. This flexibility is essential for capturing the precise flow patterns around individuals in various configurations.

3 Results and discussions

The airflow behaviour through the innovative air diffuser is depicted in the following figure (**Fig. 3**). It's evident that, owing to the diffuser's unique geometry, the airflow is characterized by convergence rather than dispersion (**Fig. 3**a,b,c). Furthermore, even beyond the converging zone, the flow maintains limited spreading (**Fig. 3**a). This observation indicates that the air diffuser exhibits low induction, aligning with the current objectives. Additionally, the turbulent kinetic energy is observed to dissipate rapidly in the vicinity of the air diffuser, signifying that the air will not transport turbulence toward the occupant, thus reducing the potential for a draft sensation (**Fig. 3d**). This favourable performance underscores the diffuser's effectiveness in maintaining occupant comfort.

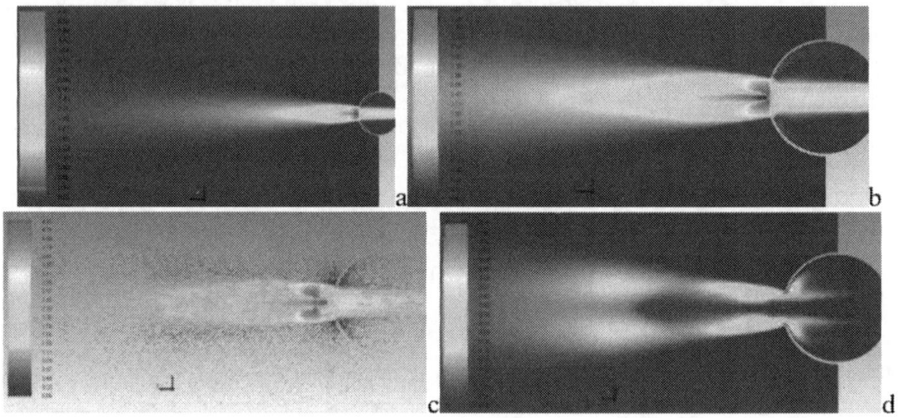

Fig. 3. Flow through the novel air diffuser a. Velocity field, b. Velocity field (close-up), c. Velocity vectors, d. Turbulent kinetic energy

An isosurface of 0.3 m/s is visible in **Fig. 4**, providing a clear depiction of how fresh air is directed toward the occupant's face.

Fig. 4. Isosurface of 0.3m/s coloured by temperature

This visualization not only demonstrates the effective airflow pattern but also highlights its role in protecting the occupant from potential pollutants present within the space. This proactive approach contributes significantly to maintaining a healthier indoor environment for the occupant and underscores its capacity to enhance occupant comfort due to low air velocity.

As Professor Fanger's work [27] explains, the air movement within enclosed spaces and its contribution to the sensation draught, it represents an undesired cooling effect on the human body caused by the air motion [27, 28]. The most widely employed quantitative model for estimating thermal comfort was proposed by Fanger [29]. In his study, subjects dressed in "standardized" clothing and engaged in "standardized" activities were exposed to various thermal conditions. In alignment with their perceived comfort, the subjects assessed this state using the ASHRAE scale [30], which comprises seven values, from very hot to cold (very hot, hot, slightly warm, comfortable, lightly cool, cool cold).

Temperature distributions in the room for the personalized ventilation system in the sagittal plane of the virtual sitting manikin (**Fig. 5**a) and in a plane passing through a lobe diffuser (**Fig. 5**b) revealed that the air provided in the proximity of the occupant face is very minimally mixed with the surrounding air.

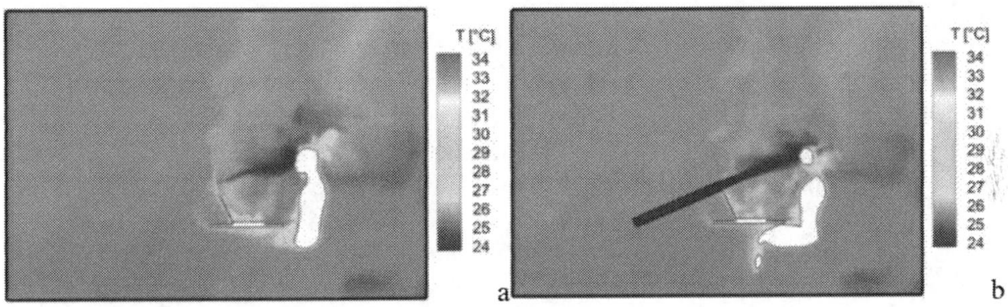

Fig. 5. Temperature distributions in the room for the personalized ventilation system a) sagittal plane, b) plane passing through an air diffuser

The same observation emerges from **Fig. 6** where it can be seen that the airflow issued from the personalized ventilation air diffuser travels almost unaffected through the breathing zone of the occupant.

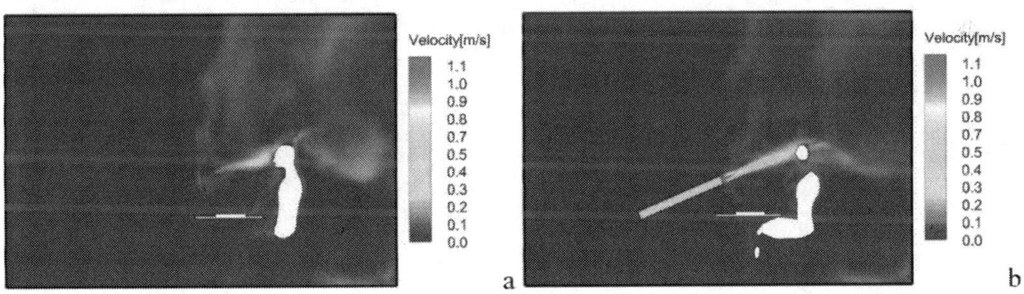

Fig. 6. Velocity distributions in the room for the personalized ventilation system a) sagittal plane, b) plane passing through an air diffuser

The velocity it's rather uniform in the proximity of the occupant face. Also, we can see that it has the capability to interact with the complex flow behaviour generated by the human body like the thermal plume of the occupant because it reaches the face of the manikin despite the above mentioned natural flow [31-33].

Regarding the draught sensation that the occupant will feel, it can be seen in **Fig. 7** that the draught rate, reaches values less than 15% in the immediate vicinity of the occupant, and aligns with the comfort sensation thresholds commonly documented in the specialized literature [34].

https://doi.org/10.1051/e3sconf/202451404004

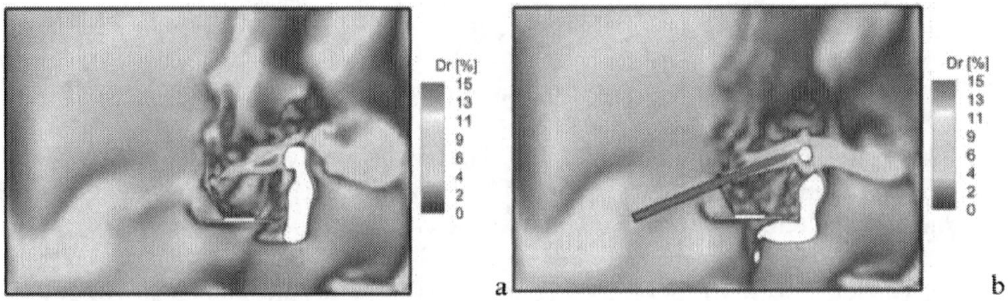

Fig. 7. Draught rate distributions in the room for the personalized ventilation system a) sagittal plane, b) plane passing through an air diffuser

Thermal comfort sensation can be estimated using Predicted Mean Vote parameter (PMV). It is a parameter used in the field of thermal comfort and indoor environmental quality. PMV is a numerical index that predicts the mean value of the thermal sensation votes of a group of people exposed to a specific thermal environment. It is used to assess and quantify how individuals are likely to perceive the thermal conditions in a given space. PMV takes into account various factors, including air temperature, mean radiant temperature, air velocity, humidity, and clothing insulation. The PMV scale typically ranges from -3 (feeling very cold) to +3 (feeling very hot), with 0 representing the point of thermal neutrality, where people are neither too cold nor too hot. PMV values above or below 0 indicate a degree of thermal discomfort [29].

In the case studied in this research it can be seen that the thermal comfort of the occupant is in the proximity of the thermal neutrality (**Fig. 8**).

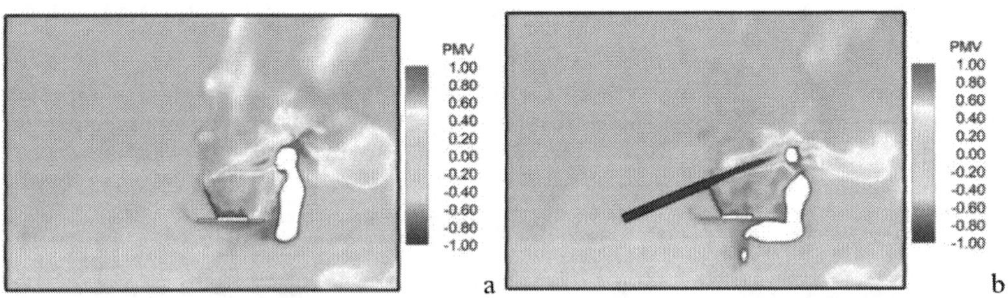

Fig. 8. PMV distributions in the room for the personalized ventilation system a) sagittal plane, b) plane passing through an air diffuser

We can see that the subject finds themselves enveloped in an environment where the Predicted Mean Vote (PMV) hovers between -1 and 1, indicative of a high level of overall thermal comfort. This range underscores that, on the whole, occupants within this setting experience conditions where they neither feel excessively cold nor uncomfortably warm. However, it's not just thermal comfort that characterizes this space; it's the simultaneous presence of fresh, clean, unpolluted air in the breathing zone that truly sets it apart.

4 Conclusions

The demand for clean, unpolluted air in office work activities cannot be overstated. It plays a pivotal role in ensuring a healthy and productive working environment. However, it's crucial not to overlook a critical aspect of this equation. While bringing cleaner air closer to the user's face is undoubtedly valuable, it alone is insufficient. For the entire concept to function effectively, the user's thermal comfort must also be satisfied.

Addressing indoor air quality is indeed a fundamental step towards creating a healthier workspace. Clean air free from contaminants, such as particulate matter and volatile organic compounds, is essential to reduce health risks, enhance cognitive performance, and ensure overall well-being. Yet, this is just one piece of the puzzle.

Thermal comfort is equally important because if a user doesn't feel comfortable, they are likely to turn off the ventilation system that brings them fresh air, negating all the benefits it offers. In other cases, employees who are too cold or too hot, those who experience draft sensation or uneven temperatures, are likely to be distracted and uncomfortable. This can hinder their concentration, reduce productivity, and even lead to health issues. Therefore, when designing systems for personalized ventilation or any indoor environmental quality improvement, achieving the right balance between clean air and thermal comfort is essential. It's not merely about purifying the air but also about creating an environment where the temperature and airflow are conducive to sustained work efficiency and occupant satisfaction. In essence, true workplace well-being is achieved when both the air we breathe and the thermal conditions we experience are harmoniously optimized.

The steerable diffuser with reduced induction by passive flow control, for use in custom ventilation systems, will deliver fresh air in proximity to the occupant's face and at the same time has been designed to solve the main problem related to this type of applications, namely the feeling of thermal discomfort of the users.

In the case of personalized ventilation systems that use a jet flow directed to the upper body, the velocity of the discharged air reaches the proximity of the face with values that exceed the accepted limits in terms of comfort. On the other hand, compared to customised ventilation systems using "laminar" air diffusers similar to displacement discharge grilles, the type of flow used also ensures an improvement in the quality of the air breathed by the user.

The proposed air diffuser delivers fresh and clean air in the proximity of the occupant but in the same time the draught rate is kept in the comfort zone, along with the PMV index.

This type of diffuser can be used in areas where it is desired to increase the quality of the air brought near the user's face and alter it with the air in the room to a lesser extent.

ACKNOWLEDGEMENT

This work was supported by a grant of the Ministry of Research, Innovation and Digitization, CCCDI - UEFISCDI, project number PN-III-P2-2.1-PED-2021-0559, within PNCDI III, - Innovative high induction air diffusers for improved indoor environmental quality in vehicles and by a grant of the Ministry of Research, Innovation and Digitization, CCCDI - UEFISCDI, project number PN-III-P2-2.1-PED-2021-2265, within PNCDI III, - Innovative seating system to reduce SARS-CoV-2 transmission on board of commercial aircrafts.

References

1. Morawska, L., et al., *Airborne particles in indoor environment of homes, schools, offices and aged care facilities: The main routes of exposure.* Environment international, 2017. **108**: p. 75-83.
2. Slezakova, K., E. de Oliveira Fernandes, and M. do Carmo Pereira, *Assessment of ultrafine particles in primary schools: Emphasis on different indoor microenvironments.* Environmental pollution, 2019. **246**: p. 885-895.
3. Ivanov, N.G., et al., *Numerical simulation of pollutant dispersion in a test ventilated room.* Journal of Physics: Conference Series, 2020. **1683**(2): p. 022075.

4. Belanger, K., et al., *Symptoms of wheeze and persistent cough in the first year of life: associations with indoor allergens, air contaminants, and maternal history of asthma.* American journal of epidemiology, 2003. **158**(3): p. 195-202.
5. Zhang, X., et al., *Environmental perceptions, mental performance, and physiological responses of people with respiratory allergies exposed to reduced Indoor Air Quality.* Indoor air, 2021. **31**(5): p. 1458-1472.
6. Franklin, B.A., R. Brook, and C.A. Pope III, *Air pollution and cardiovascular disease.* Current problems in cardiology, 2015. **40**(5): p. 207-238.
7. Cometto-Muñiz, J.E. and W.S. Cain, *Part iii. Assessing irritation: Sensory irritation: Relation to indoor air pollution.* Annals of the New York Academy of Sciences, 1992. **641**(1 Sources of In): p. 137-151.
8. Melikov, A. and J. Kaczmarczyk, *Measurement and prediction of indoor air quality using a breathing thermal manikin.* Indoor air, 2007. **17**(1): p. 50-59.
9. Georgescu, M.-R., et al. *Numerical Study of Personalized Ventilation Impact on Occupant Comfort in Enclosed Spaces*. in *2021 10th International Conference on ENERGY and ENVIRONMENT (CIEM)*. 2021. IEEE.
10. Georgescu, M.R., et al., *Personalized ventilation solutions for reducing CO2 levels in the crew quarters of the International Space Station.* Building and Environment, 2021. **204**: p. 108150.
11. Kaczmarczyk, J., et al., *Human response to five designs of personalized ventilation.* Hvac&R Research, 2006. **12**(2): p. 367-384.
12. Melikov, A.K., R. Cermak, and M. Majer, *Personalized ventilation: evaluation of different air terminal devices.* Energy and Buildings, 2002. **34**(8): p. 829-836.
13. Cermak, R. and A.K. Melikov, *Protection of occupants from exhaled infectious agents and floor material emissions in rooms with personalized and underfloor ventilation.* Hvac&R Research, 2007. **13**(1): p. 23-38.
14. Cermak, R., et al., *Performance of personalized ventilation in conjunction with mixing and displacement ventilation.* Hvac&R Research, 2006. **12**(2): p. 295-311.
15. Taheri, M., et al. *A performance assessment of an office space with displacement, personal, and natural ventilation systems*. in *Building Simulation*. 2016. Springer.
16. Zhai, Z. and I.D. Metzger, *Insights on critical parameters and conditions for personalized ventilation.* Sustainable Cities and Society, 2019. **48**: p. 101584.
17. Florin Bode, *Fluid dynamics analysis for innovative personalized ventilation diffusers for automotive and building applications*. 2011-2013, UEFISCDI: Technical University of Civil Engineering Bucharest, Romania.
18. Florin Bode, et al., *Innovative strategies of HVAC systems for high indoor environmental quality in vehicles* 2014-1016, UEFISCDI: Technical University of Cluj-Napoca.
19. Bode, F., *Innovative high induction air diffusers for improved indoor environmental quality in vehicles* 2022-2024, UEFISCDI.
20. Bode, F.I. and I. Nastase *Numerical Investigation of Very Low Reynolds Cross Orifice Jet for Personalized Ventilation Applications in Aircraft Cabins*. International Journal of Environmental Research and Public Health, 2022. **20**, DOI: 10.3390/ijerph20010740.
21. Florin Bode, et al., *A Numerical Analysis of the Air Distribution System for the Ventilation of the Crew Quarters on board of the International Space Station.* E3S Web Conf., 2018. **32**: p. 01006.
22. Angel Dogeanu, et al., *Conception of a simplified seated thermal manikin for CFD validation purposes.* Romanian Journal of Civil Engineering, 2014. **5**(1).

23. Croitoru, C., et al., *Assessment of virtual thermal manikins for thermal comfort numerical studies. Verification and validation.* E3S Web Conf., 2019. **111**: p. 02018.
24. Zasimova, M.A., E.D. Stepasheva, and N.G. Ivanov, *Effect of thermal manikin shape on thermal comfort parameters prediction uncertainties: a numerical study.* IOP Conference Series: Earth and Environmental Science, 2023. **1185**(1): p. 012040.
25. Zasimova, M.A., et al., *Evaluation of CFD-predicted thermal comfort uncertainties based on a seated thermal manikin test case.* IOP Conference Series: Earth and Environmental Science, 2023. **1185**(1): p. 012041.
26. Ansys, *Ansys Fluent theory guide.*
27. Fanger, P.O., et al., *Air turbulence and sensation of draught.* Energy and Buildings, 1988. **12**(1): p. 21-39.
28. Fanger, P.O., *The new comfort equation for indoor air quality.* Ashrae Journal, 1989. **31**(10): p. 33-38.
29. Fanger, P.O., ed. *Thermal Comfort-Analysis and Applications in Environmental Engineering*. ed. C.D.T. Press. 1970.
30. ASHRAE, *"Thermal environmental conditions for human occupancy," ANSI/ASHRAE Standard 55-2004, American Society of Heating, Refrigerating and Air-Conditioning Engineers, Atlanta, GA*. 2004.
31. Ivanov, M. and S. Mijorski, *CFD Modelling of Flow Interaction in the Breathing Zone of a Virtual Thermal Manikin.* Energy Procedia, 2017. **112**: p. 240-251.
32. Ivanov, M., *Exhaled air speed measurements of respiratory air flow, generated by ten different human subjects, under uncontrolled conditions.* E3S Web Conf., 2019. **111**: p. 02074.
33. Ivanov, M. and S. Mijorski, *Assessment of Transient CFD Techniques for Virtual Thermal Manikins' Breathing Simulations.* Environmental Processes, 2019. **6**(1): p. 241-251.
34. Kavgic, M., et al., *Analysis of thermal comfort and indoor air quality in a mechanically ventilated theatre.* Energy and Buildings, 2008. **40**(7): p. 1334-1343.

AUTHOR INDEX

EDP Sciences – Web of Conferences
17, Avenue du Hoggar
Parc d'Activité de Courtabouf
BP 112
F-91944 Les Ulis Cedex A
France

ISBN 978-1-7138-9549-7

CURRAN ASSOCIATES INC.
proceedings.com

International Conference on Entrepreneurial Competencies in a Changing World (ECCW 2021)

SHS Web of Conferences Volume 135 (2022)

Ceske Budejovice, Czech Republic
24 November 2021

edp sciences